Basics of Retaining Wall Design

A design guide for earth retaining structures

Eleventh Edition

Hugh Brooks
Civil and Structural Engineer

June, 2018

HBA PUBLICATIONS, INC.
Newport Beach, CA 92660

Basics of Retaining Wall Design

11th Edition

Hugh Brooks, PE, SE

11th Edition, first printing
Printed in the United States of America

HBA Publications, Inc.
P.O. Box 826
Corona del Mar, CA 92625
www.hbap.com

ISBN: -0-9768364-7-5

10 9 8 7 6 5 4 3 2

Disclaimer

Although it is intended that the material herein is accurate and represents good design practices, it is possible–and even likely–that errors may occur. These are code interpretations, and design practices. Other engineers may have differing views. Therefore, each of you, as engineers-of-record, must assess this information and assume responsibility for your designs. Neither the author nor HBA Publications, Inc. can assume any liability for damages resulting from use of the information in this book.

CONTENTS AT A GLANCE

1.   About Retaining Walls; Terminology
2.   Design Procedure Overview
3.   Soil Mechanics Simplified
4.   Guide to Building Codes
5.   Forces on Retaining Walls
6.   Earthquake (Seismic) Design
7.   Designing The Cantilever Wall Stem
8.   Soil Bearing and Stability
9.   Footing Design
10. Pier and Pile Foundations
11. Counterfort Retaining Walls
12. Cantilevered Tilt-up Walls
13. Wood Retaining Walls
14. Gravity Walls
15. Gabion Walls
16. Segmental Retaining Walls
17. Swimming Pool Walls
18. Pilaster Masonry Walls
19. Restrained (Non-yielding) Walls
20. Sheet Pile Walls
21. Soldier Pile Walls
22. Why Retaining Walls Fail: Effective Fixes
23. Construction Topics and Caveats
24. Design Examples

Appendix

A. Summary of Design Equations with Code References
B. Uniform System for Classification of Soil (USCS)
C. Masonry Design Data
D. Development and Lap Lengths
E. Sample Construction Notes
F. Conversion Factors
G. Reinforcing Bar Basics and US/Metric Conversions
H. Reference Bibliography
I. Notations & Symbols
J. Glossary

Index

TABLE OF CONTENTS

Why This Book?..**viii**

    The User viii
    Why It Was Written? viii
    Scope of This Book viii

Preface .... **x**

1. About Retaining Walls **1**

    Evolution of Retaining Structures 1
    A Definition 1
    Types of Earth Retaining Structures 1
    What The Terms Mean 5
    Retaining Wall Terminology 6

2. Design Procedure Overview **7**

    Design Criteria Checklist 7
    Establish the Design Criteria 8
    Basic Design Principals for Cantilevered Walls 8
    Step-by-Step Design of a Cantilevered Retaining Wall 9
    Design of a Restrained Retaining Wall 10

3. Soil Mechanics Simplified **13**

    A Soil Primer.. 13
    When is a Foundation Investigation Required? 15
    The Foundation Investigation Report 15
    The Soil Wedge Theory 16
    Explanation of Terms 17
        Angle of internal friction 17
    Soil Bearing Values 20
    Hugh's Pickle Jar Test 22
    The Precision Illusion 22

4. Guide to Building Codes **23**

5. Forces and Loads On Retaining Walls **25**

    Determining Loads and Forces 25
    Lateral Earth Pressures 25
    The Coulomb Formula 25
    The Rankine Formula 26
    Surcharge Loads 29
    Wind Loads.. 33
    Water Table Conditions 33
    Detention Ponds/Flood Walls 34
    Cascading Walls 34
    Vertical Loads 35
    Lateral Impact Loading 36

6. Seismic Design **37**

    Earthquakes – An Overview 37
    When is Seismic Design Required for Retaining Walls? 39
    Seismic Design Background 40
    Seismic for Stem Self-weight 46

7. Designing The Cantilever Wall Stem...**49**

   Basics of Stem Design 49
   Dowels from Footing into the Stem 50
   Horizontal Temperature and Shrinkage Reinforcing 51
   Key at Stem-Footing Interface 52
   Masonry Stem Design 53
   Concrete Stem Design 55

8. Soil Bearing and Stability – Cantilevered Walls **59**

   Tabulate Overturning and Resisting Moments 59
   Proportioning Pointers 59
   Overturning Moments 59
   Resisting Moments 60
   Sliding and Overturning Safety Factors 60
   Vertical Component of Active Pressure From a Sloped Backfil 61
   Determining Soil Bearing Pressure 62
   Soil Bearing Capacity 63
   Overturning Stability 64
   Sliding Resistance 64
   Footing Keys 65
   Deflection (Tilt) of Walls 66
   Global Stability 67

9. Footing Design **69**

   Basics of Footing Design 69
   Embedment of Stem Reinforcing Into Footing 69
   Toe Design 70
   Heel Design 71
   Minimum Cover for Footing Reinforcing 71
   Horizontal Temperature and Shrinkage Reinforcing 72

10. Pier and Pile Foundations **73**

   Piles, Piers, and Caissons 73
   When to Use Piles or Piers? 73
   Design Criteria 73
   Pile Design Example 74

11. Counterfort Retaining Walls **81**

   Description... 81
   Proportioning 81
   Design Overview... 81
   Designing the Wall 82
   Designing the Heel 82
   Designing the Toe... 82
   Designing the Counterfort... 83
   Stability 83

12. Cantilevered Tilt-Up Walls **85**

   Description... 85
   Construction Sequence 85
   Design Procedure 86
   Free-standing Walls 86
   Erecting the Panels 86

13. Wood Retaining Walls...**87**

14. Gravity Walls 91

Overview 91
Design procedure 91

15. Gabion and Multi-Wythe Large Walls...93

Description 93
Design 93
Foundation Pressures 94
Seismic Design 95
Gabion Walls Using Mechanically Stabilized Earth 95

16. Segmental Retaining Walls (SRWs) 97

Overview 97
Gravity Wall Design 98
    Check Lateral Soil Pressures 100
    The Coulomb Equation 100
    Check Inter-Block Shear 100
    Check Sliding 100
    Check Overturning 100
    Check Soil Bearing Pressure 101
    Soil Bearing Capacity 101
    Seismic Design 101
    Gravity Walls 101
Geogrid Wall Design.. 102
    Construction Sequence 103
    About Geogrids 103
    Gather Design Criteria 104
    Select Masonry Units 104
    Determine Lateral Soil Pressures 105
    Select Geogrid 105
    Determine Geogrid Embedment 106
    Determine Depth of Reinforced Soil (total base width) 107
    Check Overturning 107
    CheckSliding at Lowest Geogrid Layer 108
    Check Sliding at Base 109
    Check Soil Bearing Pressure 109
    Soil Bearing Capacity 109
    Seismic Design 110
    Building Codes & Standards 111
Getting Help 112

17. Swimming Pool Wall Design 113

18. Pilaster Masonry Walls 115

Description 115
Filler Wall Design 115
Pilaster Design 115
Footing Design 116

19. Restrained (Non-Ylielding) Walls 117

Description 117
Dual Function Walls 117
At Rest Active Soil Pressure 117
Seismic Force on Non-Yielding (Restrained) Walls 118

20. Sheet Pile Walls 119

Description 119
Design Procedure 119
References 120

21. Soldier Beam Walls 121

Description 121
Design Procedures 121

22. Why Retaining Walls Fail and Cost - Effective Fixes 125

23. Construction Topics and Caveats 131

Horizontal Control Joints 131
Drainage 131
Backfill 131
Compaction 132
Inspections 132
The Geotechnical Investigation 132
Forensic Investigations 132

24. Retaining Wall Design Examples 133

Appendix 228

A. Summary of Design Equations with Code References 228
B. Unified Soil Classification System (USCS) 230
C. Masonry Design Data 231
D. Development and Lap Lengths 232
E. Sample Construction Notes 233
F. Conversion Factors 234
G. Reinforcing Bar and US/Metric Conversions 235
H. Reference Bibliography 236
I. Notations & Symbols 237
J. Glossary 239

INDEX 249

"Theory predicts how soil <u>should</u> behave,
. . . but soil doesn't always comply"

J. P. Nielsen
Geotechnical Engineer

WHY THIS BOOK?

For The User

This book is intended to cover and explain design practices and building code requirements for the design of earth retaining structures. It is for both the practicing engineer who has become a bit rusty on this complex subject, and for the interested engineering student who has already acquired a basic knowledge of statics, soil mechanics, and the design of simple masonry and concrete structures. Design review agencies will also find it a useful reference.

Because building codes now apply to nearly all design procedures an important feature of this book are current code references throughout. These need to be updated in succeeding editions.

Why It Was Written

My objective in each edition was to cover basic design principles and practices in a concise, readable, manner. This book is not an in-depth treatment of the design of retaining structures. Earth retaining structures and soil mechanics are far too complex a subject to treat in a single concise volume. There are dozens of foundation engineering texts and countless technical papers available for review, and of course there is the Internet. However, finding what you need is time consuming; hence this compendium. The challenge was to decide what to put in and what to leave out and to put in the most helpful things a designer – and an advanced engineering student -- need to know to design most types of earth retaining structures. Surely there will be omissions and errors, but the intent is that you will find this book helpful in your practice.

Scope of This Book

This book treats design practices and current code requirements for most types of earth retaining structures: conventional cantilevered retaining walls: restrained (basement) walls, gravity walls, and segmental retaining walls both gravity and with geogrids. Other topics include sheet pile walls, tilt-up retaining walls, soldier pile walls, gabion walls, counterfort walls, pilaster walls and walls with pile or pier foundations. A review of basic soil mechanics and seismic design are also included in this book and the appendix offers useful information including a glossary of terminology.

Your feedback for comments, suggestions and corrections will be welcome.

Contact us at hbrooks@hbap.com.

Hugh Brooks, PE, SE
June 2018

THIS PAGE LEFT BLANK FOR NOTES

PREFACE TO THE ELEVENTH EDITION

This is the eleventh edition in a series which began in 1996 as a modest companion manual to accompany my work developing Retain Pro software (now a division of Enercalc.com). In successive editions it has grown from 93 pages to 250 in this current edition, with many thousands of copies in print.

This edition, like its predecessors, becomes necessary because of three-year building code cycles change – (seismic design requirements for example) – and suggestions we receive for topics to be added or expanded. To respond, successive editions have been updated, corrections made and additional materials added for this edition. All design examples have been reviewed, corrections made where needed and explanations added. All code references have been updated and seismic design has been both updated and expanded.

This edition primarily references these codes: ASCE 7-16, ACI 318-14, and IBC 2018. At this writing, (5/2018) the latter code has yet to be widely adopted and referenced. However, most references relating to earth retaining structures are unchanged from IBC 2015 or explained if changed.

My appreciation to the many of you who have offered valuable suggestions, corrected errors, read portions of the draft, offered informative articles and excerpts from technical papers. My thanks to each of you. We will all benefit from your input. And a special acknowledgment to my co-author of the 10th edition, John Nielsen, civil and geotechnical engineer, for his many valuable contributions and insightful suggestions.

I trust this eleventh edition will be helpful in your practice and a valuable reference.

As always, your comments and suggestions are welcomed! hbrooks@hbap.com.

Hugh Brooks, P.E., S.E.

THIS PAGE LEFT BLANK FOR NOTES

Evolution of Retaining Structures

In the year one-million BC, or thereabouts, an anonymous man, or woman, laid a row of stones atop another row to keep soil from sliding into their camp. Thus was constructed an early retaining wall, and we've been keeping soil in place ever since…… with increasingly better methods and understanding.

The early engineers in the ancient cultures of Egypt, Greece, Rome and the Mayans were masters at invention and experimentation, learning primarily through intuition and trial-and-error what worked and what didn't We marvel at their achievements. Even the most casual observer looks in wonder at the magnificent structures they created and have stood for thousands of years – including countless retaining walls. With great skill they cut, shaped, and set stone with such precision that the joints were paper thin. Reinforced concrete would not be developed for a thousand years, but they used what they had, and learned how to do it better with each succeeding structure. Consider the Great Wall of China, for example, where transverse bamboo poles were used to tie the walls together – a forerunner of today's "mechanically stabilized earth". Those early engineers also discovered that by battering a wall so that it leaned slightly backward the lateral pressure was relieved and the height could be extended – an intuitive understanding of the soil wedge theory. Any student of ancient construction methods is awed by the ingenuity and accomplishments of those early engineers.

Major advances in understanding how retaining walls work and how soil generates forces against walls appeared in the 18th and 19th centuries with the work of French engineer Charles Coulomb 1776, who is better remembered for his work on electricity, and later by William Rankine in 1857. Today, their equations are familiar to most civil engineers. A significant body of work was the introduction of soil mechanics as a science through the pioneering work of Karl Terrzaghi in the 1920s.

Indeed, soil mechanics and the design of retaining structures have advanced dramatically in recent decades giving us new design concepts, a better understanding of soil behavior, and hopefully safer and more economical designs.

A Definition:

A retaining wall is any constructed wall that restrains soil or other material at locations having an abrupt change in elevation.

Types of Retaining Structures

There are many types of structures used to retain soil and other materials. Listed below are the types of earth retaining structures generally used today. The design of these will be discussed in later chapters.

Cantilevered retaining walls

These walls which retain earth by a wall cantilevering up from a footing are the most common type of retaining walls in use today. These walls are classified as "yielding" as they are free to rotate (about the foundation) because of the lack of any lateral restraint above its base.

Cantilevered retaining walls are generally made of masonry or concrete, or both, but can also take other forms as will be described.

Types of Cantilevered Retaining Walls Include:

Masonry or concrete walls

The stem of a masonry wall is usually constructed of either 8″ or 12″ deep concrete masonry block units. The cells are partially or solid grouted, and are vertically reinforced. An eight-inch block is generally adequate to retain up to about six feet, and a twelve-inch block up to ten to twelve feet of soil.

The stems of a concrete wall must be formed, and can be tapered for economy, usually with the taper on the inside (earth side) to present a vertical exposed face.

Above the base where higher strength is required, hybrid walls, consisting of both concrete and masonry, can also be constructed using formed concrete that extends some height then changed to masonry to the top of the wall.

A variation of masonry cantilever walls consisting of spaced vertical pilasters (usually of square masonry units) with in-filled walls of lesser thickness, usually 6″ masonry. The pilasters cantilever up from the footing and are usually spaced from four to eight feet on center. These walls are usually used where lower walls are needed – under about six feet high.

Counterfort retaining walls

Counterfort cantilevered retaining walls incorporate wing walls projecting upward from the heel of the footing into the stem. The thickness of the stem between counterforts is thinner (than for cantilevered walls) and spans horizontally, as a beam, between the counterfort (wing) walls. The counterforts act as cantilevered elements and are structurally efficient because the counterforts are tapered down to a wider base at the heel where moments are the highest. The high cost of forming the counterforts and the infill stem walls make such walls usually not practical for walls less than about 16 feet high. See Figure 1.1.

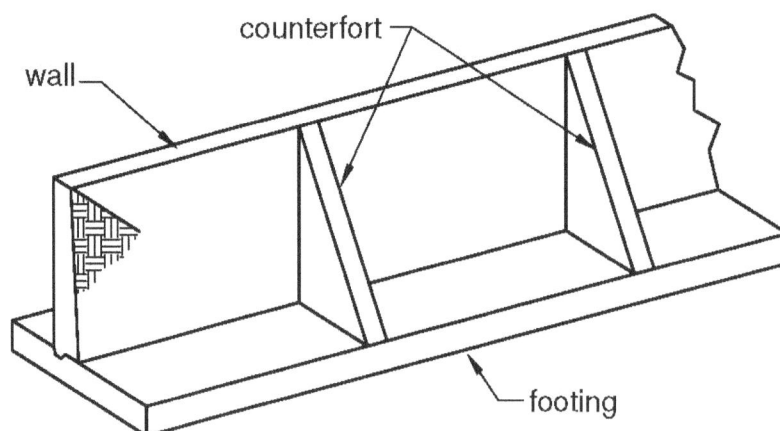

Figure 1.1 Counterfort retaining wall

Buttress retaining walls

These are similar to counterfort walls, but the wings project from the outside face of the wall. Such walls are generally used in those cases where property line limitations on the earth retention side do not allow space for the large heel of a traditional cantilevered retaining wall.

Gravity retaining walls

This type of wall depends upon the dead load mass of the wall for stability rather than cantilevering from a foundation.

Stacked and mortar-bonded stone, rubble, or rock walls

These are usually gravity walls relegated to landscaping features with retaining less than about four feet high. Engineering for such walls is limited, or none at all, and rules-of-thumb prevail (such as a retained height not more than two or three times the base width). Higher walls need engineering to evaluate overturning, sliding, soil bearing and to verify that flexural tension does not exist within the wall (or only as allowed by code for material used) because these walls are generally unreinforced.

Gabion or crib walls

A gabion wall is a type of gravity wall whereby stones or rubble are placed within wire fabric baskets. Crib walls are a variation of the gabion method whereby mostly steel bins are filled with stone or rubble. Another variation is to stack a grillage of timbers and fill the interior with earth or rubble. Precast concrete cribs are also widely used to form crib walls.

Wood retaining walls

Wood is commonly used for low height retaining walls. Wood retaining walls usually consist of laterally spaced wood posts embedded into the soil, preferably into a drilled hole with the posts encased in lean concrete. Horizontal planks span between the upward cantilevering posts. Pressure treated wood is used, but even with treatment deterioration is a disadvantage, and wood walls are generally limited to low walls because height is limited by size and strength of the posts. Railroad ties are commonly used for both posts and lagging.

Tilt-up concrete retaining walls

Tilt-up concrete construction has been successfully used for retaining walls, either cantilevered or restrained at the top. These site-cast panels are set on isolated concrete pads placed at panel ends, with reinforcing projecting out from the bottom of each panel. A continuous concrete footing is then cast under the wall to complete the construction. Tilt-up walls are economical for higher walls, but sufficient space is needed to cast the panels.

Segmental retaining walls (SRWs)

Many manufacturers offer various systems of stacked segmental concrete units, steel bins, or other devices that retain soil by stacking individual components. SRW's are essentially gravity retaining walls. Most are patented systems that are typically battered (sloped backward), primarily to reduce lateral soil pressure, thus requiring a minimal foundation. Reinforced concrete footings, steel reinforcing, or mortar are not used. Stability of SRW gravity walls depends solely upon the dead weight resisting moment exceeding the lateral soil pressure overturning moment. To attain greater heights – up to 40 feet and more – SRW's also utilize mechanically stabilized earth (MSE), also called reinforced earth, whereby geosynthetic fabric layers are placed in successive horizontal layers of the backfill to achieve an integral soil mass that increases resistance to overturning and horizontal sliding. A variety of facing block configurations and surface colors and textures are available from many manufacturers.

Bridge abutments

These support the end of a bridge and retain the earth embankment leading to the bridge. Bridge abutments usually have angled wing walls of descending height to accommodate the side slope of the embankment. Abutments are designed as cantilever walls, with girder bearing support free to slide at one end to accommodate horizontal expansion movement of the bridge deck. Design requirements for bridge structures are usually governed by the code requirements of the American Association of State Highway and Transportation Officials (AASHTO) and state Departments of Transportation (DOTs) such as California's CalTrans.

Sheet pile and bulkhead walls

These are generally waterfront structures such as at docks and wharves, but steel sheet piling is also used for temporary shoring on construction sites. Steel sheet units configured for stiffness or concrete panels are driven into the soil to provide lateral support below the base of the excavation or the dredge line. Sheet pile walls cantilever upward to retain earth or are restrained at or near the top by either a slab-on-grade or tiebacks.

Restrained (Non-yielding) retaining walls

Also called "basement walls" (for residential and light commercial conditions), or "tie-back" walls. These walls are distinguished by having lateral support at or near the top, thereby with less or no dependence for fixity at the foundation. Technically, these walls are classified as "non-yielding" walls because such walls cannot move laterally at the top, as opposed to cantilevered (yielding) walls. Such walls are usually designed as "pin connected" both at the top and bottom. The earth pressure creates a positive moment in the wall, which requires reinforcing on the front of the wall, that is, the side opposite the retained soil. In some cases it may be cost effective to fix the base of the wall to the footing to reduce both the bending in the wall and restraining force required at the top support.

Footings for these walls are usually designed for vertical loads only. However, it is often desirable to design the lower portion of a basement wall as a cantilevered retaining wall with fixity at the footing so that backfill can be safely placed to avoid bracing the wall, or waiting until the lateral restraint at the top is in place, such as a floor diaphragm. Note that conventional wood floors framed into the top of a basement wall may not provide a sufficient stiffness to allow for the restrained case,

Anchored (tieback) walls

Anchors or tiebacks are often used for higher walls where a cantilevered wall may not be economical. Restraint is achieved by drilling holes and grouting inclined steel rods as anchors into the zone of earth behind the wall beyond the theoretical failure plane in the backfill. The anchors can be placed at several tiers for higher walls, and can be post-tensioned rods grouted into drilled holes, or non-tensioned rods grouted into the drilled holes. The latter are also known as soil nails.

What the Terms Mean:

Backfill: Soil placed behind a wall.

Backfill slope: Often the backfill slopes upward from the back face of the wall. The slope is usually expressed as a ratio of horizontal to vertical (e.g. 2:1).

Batter: Slope of the face of the stem from a vertical plane, usually on the inside (earth) face.

Dowels: Reinforcing steel placed in the footing and bent up into the stem a distance at least equal to the required development length of each reinforcing bar.

Footing (or foundation): That part of the structure below the stem that supports and transmits vertical and horizontal forces into the soil below.

Footing key: A deepened portion of the footing usually placed directly under the stem to provide increased sliding resistance.

Grade: The surface of the soil or paving; can refer to either side of the wall.

Heel: That portion of the footing extending behind the wall (under the retained soil).

Horizontal temperature/shrinkage reinforcing: Longitudinal horizontal reinforcing usually placed in both faces of the stem and used primarily to control cracking from shrinkage or temperature changes.

Keyway: A horizontal slot located at the base of the stem and cast into the footing for greater shear resistance.

Principal reinforcing: Reinforcing used to resist bending in the stem.

Retained height: The height of the earth to be retained is generally measured upward from the top of the footing.

Stem: The vertical wall above the foundation.

Surcharge: Any load placed in or on top of the soil, either in front of or behind the wall.

Toe: That portion of footing which extends in front of the front face of the stem (away from the retained earth).

Weep holes: Holes provided at the base of the stem for drainage. Weep holes usually have gravel or crushed rock behind the openings to act as a sieve and prevent clogging. Poor drainage of weep holes is the result of weep holes becoming clogged with weeds, thereby increasing the lateral pressure against the wall. Unless properly designed and maintained, weep holes seldom "weep". Alternatively, perforated pipe surrounded with gravel and encased within a geotextile can be used to provide drainage of the backfill.

Cantilevered Retaining Wall Terminology

Cantilevered retaining walls have unique descriptive terminology as illustrated below:

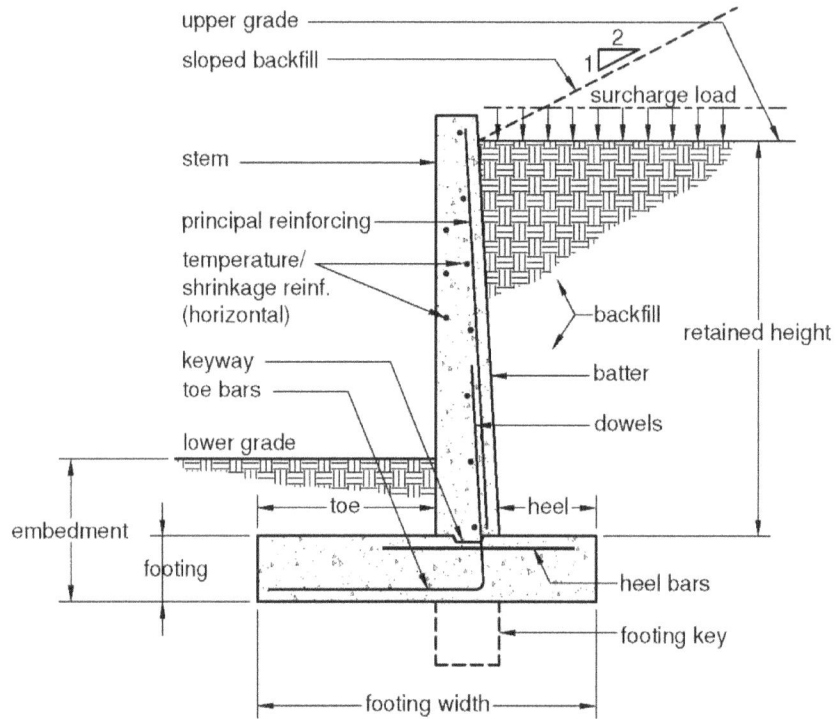

Figure 1.2 Retaining wall terminology

The Four Primary Concerns for the Design of Nearly any Retaining Wall are:

1. That it has an acceptable Factor of Safety with respect to overturning.
2. That it has an acceptable Factor of Safety with respect to sliding.
3. That the allowable soil bearing pressures are not exceeded.
4. That the stresses within the components (stem and footing) are within Code allowable limits to adequately resist imposed vertical and lateral loads. It is equally important that it is constructed according to the design.

Design Criteria Checklist

Before establishing specific design criteria, the following checklist should be used before starting your design:

- What Building Codes are applicable?
- Do I have the correct retained height for all of my wall conditions?
- Is there a property line condition I need to know about?
- Is there a fence on top of the wall, or does the wall extend above the retained height? (exposure to wind)
- How deep must the bottom of the footing be (frost considerations)?
- How will I assure that the backfill will be drained?
- Will there be any axial loads on top of the wall? If so, their eccentricity?
- What about surcharges behind the wall, such as parking, trucks, etc.?
- If the wall extends above the higher grade, and is a parking area, is there an impact bumper load?
- What is the slope of the backfill? Level?
- Is there a water table I need to consider?
- Is a seismic design required?
- Are there any adjacent footing loads affecting my design?
- Should the stem be concrete or masonry, or a combination of the two?
- How high is the grade on the toe side, above the top of the footing?
- Is there a slab in front of the wall to restrain sliding or provisions to prevent erosion of soil along the toe?
- Is there lateral restraint at the top of the wall (if so, it's not truly a cantilevered wall and requires a different design)?
- Do I have a soil investigation report or other substantiation for soil properties: active pressure, passive pressure, allowable bearing pressure, sliding coefficient, soil density, and other items I need to consider?
- Also consider whether a cantilevered retaining wall is the right solution. If the height of the wall is over about 16 feet, perhaps a tieback wall would be more economical (caution: be sure your client has the right to install tiebacks. If the wall is on a property line, there is obviously a problem). Perhaps a buttressed or counterfort wall would be better for high walls, or using

precast panels, or tilt-up to overcome construction constraints imposed by a restrictive rear property line.

- Lastly, determine how many conditions for which you will need a design. Perhaps the same retained height has several different backfill slopes, say, from level to 2:1. Here you need to use a little judgment in determining the number of cases. Usually you don't design for less than two-foot height increments, unless there are different surcharges or other conditions. To design for one-foot height increments is not only tedious, but doesn't save that much material cost. On the other hand, if the retained height along the length of a wall varies from, say, four feet to 12 feet, you would not want to specify the 12-foot design throughout. In this case, you would probably design for 4', and 6', 8', 10', 12'. You rarely "design" a wall less than 4 feet high, just use a little judgment—unless there is a steep backfill slope or large surcharges, in which case it should be properly designed.

Establish Design Criteria

The following information will be needed before starting your design. The values shown in parentheses are only given to illustrate those values frequently used.

Retained heights
Embedment depth of footing required below grade – See geotechnical report
* Allowable soil pressure (1,000 psf to 3,000 psf)
* Passive pressure (150 to 350 pcf)
* Active earth pressure (30 pcf to 55 pcf)
* Coefficient of friction (.25 to .40)
Backfill slope (don't exceed about 2:1 horizontal: vertical unless approved by the geotechnical engineer)
Axial loads on stem
Surcharge loads
Wind, if applicable
* Seismic criteria if applicable
* Soil density (110 to 130 pcf)
Concrete and masonry allowable stresses (usually used values in parentheses – also see Appendix I – Notations & Symbols).
f'_c (2,000 psi to 4,000 psi)
f_y (60,000 psi)
f_s (32,000 psi)
f'_m (1,500 psi)
f_r (145 psi to 178 psi, strength design)

* These values are usually given in the geotechnical report.
When you have gathered this information, you're ready to start.

Basic Design Principals for Cantilevered Walls

Stability requires that a cantilevered retaining wall resists both overturning and sliding, and material stresses including the allowable soil bearing pressure that must be within acceptable values.

To resist forces tending to overturn the wall (primarily the lateral earth pressure against the back of the wall), the wall must have sufficient weight, including the soil above the footing, such that the resisting moments are greater than the overturning moments. The safety factor for overturning should be at least 1.5 – some codes require more.

To resist sliding, the weight of the wall plus the weight of the soil above the footing plus vertical loads on the wall and any permanent surcharges multiplied by the coefficient of friction between the foundation soil and the bottom of the footing, plus the passive pressure resistance force at the front of the wall, must be sufficient to resist the lateral force on the wall. The recommended safety factor against sliding is 1.5. If the soil is cohesive, the coefficient of friction is replaced by the adhesive bond (see page 65) of the cohesion between the footing and soil, in psf.

The stem must be designed to resist the bending caused by lateral earth pressures, including the effect of surcharges placed behind the wall, seismic or wind if applicable, impact loads, or axial loads acting eccentrically on the wall. The maximum bending and shear stresses in a cantilevered wall will be at the bottom of the stem. Each of these subjects will be discussed later.

Figure 2.1 is a free-body force diagram illustrating stability forces on a cantilevered wall.

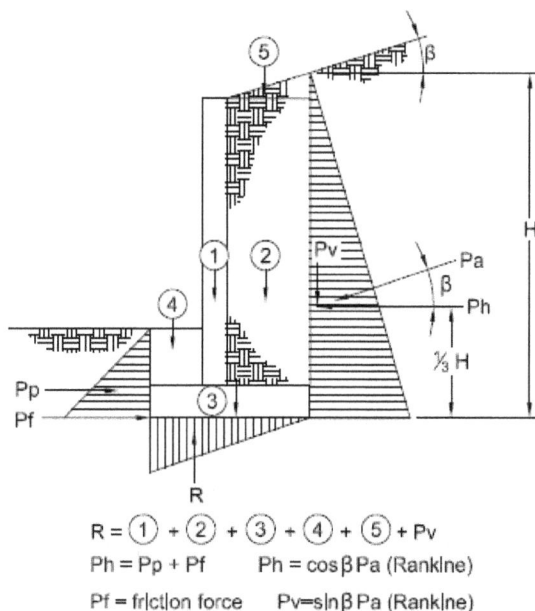

$$R = \text{①} + \text{②} + \text{③} + \text{④} + \text{⑤} + Pv$$
$$Ph = Pp + Pf \qquad Ph = \cos\beta\, Pa \ (\text{Rankine})$$
$$Pf = \text{friction force} \qquad Pv = \sin\beta\, Pa \ (\text{Rankine})$$

Figure 2.1 Free-body of cantilevered retaining wall

Step-by-Step Design of a Cantilevered Retaining Wall

The design usually follows this order:

1. Establish all design criteria based upon applicable building codes. (See checklist above).

2. Compute all applied loads, soil pressures, seismic, wind, axial, surcharges, impact, or any others.

3. Consider Load Combinations. See below.

4. Design the stem. This is usually an iterative procedure. Start at the bottom of the stem where moments and shears are maximum. Then, for economy, check several feet up the stem (such as at the top of the development length of the dowels projecting from the footing (See page 50) to determine if the stem bar size can be reduced or alternate stem bars dropped. Check dowel embedment depth into the footing assuming a 90° bend (hooked bar). The thickness of the stem may vary, top to bottom. The minimum top thickness for reinforced concrete walls is usually 6- inches to properly place the concrete; 8-inches at the bottom.

5. Compute overturning moments, calculated about the front (toe) bottom edge of the footing. For a trial, assume the footing width, to be about 1/2 to 2/3 the height of the wall, with 1/3 being at the toe.

6. Compute resisting moments based upon the assumed footing width, calculated about the front edge of the footing.

7. An overturning factor of safety of at least 1.5 is considered standard of practice.

8. Check sliding. A factor of safety with respect to sliding of 1.5 or more is standard. A key or adjusting the footing depth may be required to achieve an accepted factor of safety with respect to sliding.

9. Based upon an acceptable factor of safety against overturning, calculate the eccentricity of the total vertical load. Is it within or outside the middle-third of the footing width?

10. Calculate the soil pressure at the toe and heel. If the eccentricity, e, is > B/6 (B = width of footing) it will be outside the middle third of the footing width (not recommended!), and because there cannot be tension between the footing and soil, a triangular pressure distribution will be the result. Consult with the project geotechnical engineer if this condition cannot be avoided, as it will result in a substantially lowered allowable soil bearing pressure.

11. Design footing for moments and shears by selecting reinforcing.

12. Check and review. Have all geotechnical report requirements been met?

13. Place a note on the structural sheets and on the structural calculations indicating that the backfill is to be placed and compacted in accordance with the geotechnical report.

14. Review the construction drawings and specifications for conformance with the design.

Step-By-Step Design of a Restrained Retaining Wall

Similar to the above with some additional steps (italicized):

1. Establish all design criteria based upon applicable building codes. (See checklist above).

2. Compute all applied loads (at-rest earth pressures, seismic, wind, axial, surcharges, impact, or any others. *Select "height" to lateral restraint.*

3. *Select restraint – level and base of stem design assumptions: pinned - pinned; pinned fixed; or fixed - fixed. Then based on statics determine the reactions at the top and at the base of the wall.*

4. *If a floor slab is present at the top of the footing, check its adequacy to sustain this lateral sliding force.*

5. Design the stem. *If the stem is assumed pinned at the base and at the top, the maximum moment will be a positive moment near mid-height—select stem material, design thickness, and reinforcing for that location. Usually the same material (concrete or masonry) and thickness will be used for the full height. Some degree of "fixity" is likely at the top of the wall even with a pinned "design."*

6. Design the footing. *If the stem is assumed fixed at the base check the soil pressure (check Items 8 and 9 as above) and design for the moments and shears, and select reinforcing. If the stem is assumed pinned at the footing interface, try to center the footing under the wall to prevent eccentricity. If there is eccentricity check reinforcing at stem-footing interface to resist that moment because if it exceeds the moment because of the eccentricity the soil*

pressure will not be uniform Check embedment depth into the footing assuming a 90° bend (hooked bar).

7. Check sliding. *If a restraining floor slab is not present, a key or adjusting the footing width or depth may be required.*

8. Check and review. *Have all soil report requirements been met?*

9. Review the construction drawing for conformance with your design.

All these topics will be discussed later.

Load Combinations:

Load Combinations for Load and Resistance Factor Design (LRFD), also termed Strength Design, which are applicable to retaining walls per IBC 2018, 1605.2 are shown below.

$$DL = 1.2 \quad \text{(Dead Load)}$$
$$LL = 1.6 \quad \text{(Live Load)}$$
$$H = 1.6 \quad \text{(Earth pressure)}$$
$$W = 1.6 \quad \text{(Wind)}$$
$$E = 1.0 \quad \text{(Seismic)}$$

For Allowable Stress Design (ASD) use 1.0 for all except 0.70 for seismic and 0.6 for wind but not used simultaneously. See IBC 2018,1605.3.1.

Load combinations for footing strength are typically strength-level combinations, because footings are typically concrete.

Load combinations for concrete stem design strength are typically strength-level combinations. Masonry Stems are usually Allowable Stress Design (ASD) combinations.

Safety Factors

The safety factor for retaining walls to resist overturning and sliding is 1.5. When earthquake loads are included the safety factor can be 1.1.

The above Load Combinations do not apply to the safety factors. Instead, base the seismic design to 0.7 earthquake force and 1.0 for other nominal forces.
Refer to IBC 2018, 1807.2.3. For Risk Factor adjustments the IBC 2018, Table 1604.5.

A Soil Primer

The mantle of our earth is composed of water, rock and soil. It is the soil or rock that supports our structures. We need to understand what soil is, how it behaves, and the properties we need for design. Soil is a collective term for any mixture of sand, silt, or clay. Soil is not "dirt", which we sweep off the floor and wash from our clothes. Dirt is a colloquial term contractors often use, such as "We underestimated the fill quantity and need to import 200 more yards of dirt (a "yard" in that terminology means one cubic yard).

Soil is the result of the decomposition of rock. Rocks decompose by weathering, freezing and thawing, by crushing and grinding along earthquake faults, along planes of failure in landslides, by the overland movement of glaciers, the tumbling action of rivers and streams, and from the corrosive inorganic acids present in the atmosphere and derived from plants. Additionally, we must add heat, temperature changes and pressures within the earth.

Before the mid-1920s, determining how large a footing was needed to support a structure was rudimentary. It consisted primarily of driving rods into the soil and observing the resistance, auger borings, test pits, and usually load testing a small area and observing tolerable deformations from which a footing could be safely sized. Recommended bearing capacities were published in the handbooks of the day. For instance, the 1916 *New York Building Code* listed capacity of various soils. An example: "Sand and clay mixed or in layers" allowed "2 tons per square foot".

A pioneer to advance soil behavior to a science was Karl Terzaghi (1883-1963) who in 1925 published *Erdbaumechanik*, which loosely translates to mechanics of soil in construction, followed in 1926 by *Principles of Soil Mechanics*. Later, in 1948, he and Peck authored the classic *Soil Mechanics in Engineering Practice*. His studies were based upon application of the theory of elasticity to mass materials. His work, and that of others, resulted in the term *soil mechanics* which later when combined with principles of geology, evolved into *geotechnical engineering*.

Moving ahead to today, types of soil, sand, silt, or clay, primarily are classified by particle size and the composition of the mixture. The distribution of grain size in a soil sample is determined by a grain size distribution analysis. For example, in a sieve test a sample is passed through successively smaller sieves, and the amount by weight retained on each sieve is noted as a percent of the total. With this information the geotechnical engineer can classify the soil per the most-used *Uniform System for Classification of Soil* (USCS) that is reproduced in Appendix B. Sieve sizes use a numbering system where the number indicates the number of openings per inch. For example, a #4 sieve has four openings per inch, or ¼" each, and a #200 sieve has 200 openings per inch, and so forth. Grain size-distributions for soil particles finer than the #200 sieve are determined by hydrometer techniques.

Some common designations of soil are:

Boulders	>	12″
Cobbles	>	3″ < 12″
Gravel	>	#4 sieve < 3″
Sand	>	#200 sieve < #4 sieve
Silt	<	#200 sieve
Clay	<	0.005 to 0.002 mm

There are other classifications systems, such as the AASHTO system (American Association of State and Highway Transportation Officials), but the USCS classification is most often referred to in the foundation investigation reports you will read.

Soil is further classified as being cohesive, non-cohesive, or somewhere in between.

Cohesive soil derives its strength primarily from the cohesive bond between particles. Examples include fine-grained silts and clays.

Non-cohesive, or granular soil, derives its strength from inter-particle friction between grains. Sand and gravel are examples of non-cohesive soil. Non-cohesive soil is the type usually assumed for analysis of pressures against a retaining wall.

Expansive soil usually consists of clay, but some silts are also expansive. Expansive soil can lift footings depending upon the amount of water present or shrink upon drying. Some clays are highly expansive changing in volume with changes in water content. Such swelling can cause considerable pressure on retaining structures. It is for this reason that clay backfill should be avoided, and if the site contains expansive soil, the geotechnical engineer will recommend measures to minimize its effect, mainly by removal and replacement with suitable material. It is important that water not be allowed to penetrate into expansive soil.

Frost line is a term used in colder climates in the northern US, whereby upper portions of the ground may freeze seasonally or permanently, with frost depths ranging from a few inches to eight feet or more. To prevent the added pressure of swelling because of freezing, foundations should be placed below the frost line. The geotechnical engineer and applicable building codes will address this local concern. In areas where the ground is permanently frozen to a great depth, such as Alaska, local expertise and experience will apply.

Bearing capacity of a soil is an estimate of its capability to support a vertical load in compression but does not expresss the amount of settlement that may occur. The shearing strength of the soil is the controlling factor for determining its bearing capacity. The shear between particles can be either frictional resistance (sliding friction between particles) or in the case of a clayey soil, cohesion and perhaps interparticle friction. Sandy soil requires confinement to develop shear strength, as for example a lack of confinement is illustrated when you step on sand at the beach you will notice the sand displaces sideways under your feet. This illustrates the lack of frictional forces at work.

When soil samples (cores retrieved from drilling) are taken to the laboratory for testing, the geotechnical engineer will calculate the bearing capacity of the particular soil by determining its angle of internal friction, ϕ, its unit cohesion, c, and its unit weight.

Most soil mechanics texts will thoroughly cover the several types of shear tests available to the geotechnical engineer.

The basic equation for shear resistance developed along a plane of rupture is:

$$s = c + p \tan \phi$$

s = shear strength; p = effective normal stress and c = cohesion, both usually expressed in psf; and ϕ = effective angle of internal friction.

When is a Foundation Investigation Required?

To determine if a site is suitable for the loads to be imposed, and to issue design recommendations, a geotechnical report is usually required. This requirement is in section 1803 of IBC '18. Refer to Chapter 4 for information on Building Codes. The local building official may have the authority to waive an investigation report if the soil is reasonably well known or a report was prepared for a nearby site. When a report is required the requirements are specified in IBC '18, section 1803.2. That section specifies the soil classification be per ASTM D 2487, which is the USCS system. These reports must be prepared by a "registered design professional", which in practice is a state licensed civil engineer. However, there is a trend towards the term "geotechnical engineer". This license is designated by "G.E." and is earned by a rigorous examination and experience requirements.

The Foundation Investigation Report

When a soil or geotechnical report is provided the designer must review it carefully. If any questions, concerns or alternate procedures arise, contact the geotechnical engineer responsible for the report. The report contains recommendations for design soil bearing pressures and, if applicable, site preparation procedures if fill or poor soil is present, and other recommendations that may impact structural design, such as predicted settlement. Results of soil bores (usually a minimum of three per site) are displayed along with layered soil classification based upon USCS standards. There may be geological information which will comment on nearby faults, anticipated levels of seismic accelerations, the potential for landslides, the presence of potentially expansive soil and other potential concerns.

List of soil tests usually performed to develop a Soil Report:

Tests that suggest probable soil behavior
- Moisture content
- Wet and Dry Densities
- Percent saturation
- Atterberg Limits
    Shrinkage Limit
    Plastic Limit
    Plasticity Index
- Grain-Size Distribution
- Expansion Index

List of Tests that Report on Soil Strength and Compressibility:

- Direct Shear
- Unconfined Compression
- Triaxial
- Consolidation
- Expansion
- Sensitivity
- Moisture – Density (Compaction Curves)
- R – Value or CBR

Information that may be included in a report:

- Soil classification
- Allowable soil bearing value
- Adjustments in soil bearing for increase in width and depth of footings.
- Passive soil pressure
- Active soil pressure for various backfill slopes.
- Coefficient of friction (concrete to soil).
- At-rest active pressure for restrained (non-yielding) walls.
- Presence of ground water
- Liquefaction potential
- Slope stability analyses
- Seismicity (peak ground acceleration, proximity to faults, etc.)
- Presence of fill and site grading requirements.
- Other constraints that the designer should be aware of.

The Soil Wedge Theory for Retaining Walls

How much force does the retained earth impose on a retaining wall? One of the early investigations of this problem was reported in a 1729 publication by French engineer Bernard Belidor. He started with a simple premise: If a wall retaining soil was suddenly removed, the soil behind it would slide down, slipping along a plane he assumed was inclined at 45°. He reasoned, and solved by simple statics, that if the plane was without friction, the horizontal force against the wall would be equal to the weight of the "wedge" of soil. He then assumed a 0.5 friction factor along the sloping plane, which then halved the lateral force; the lateral force was about one-half the weight of the soil wedge.

Later in the 1770's, French engineer Charles Coulomb further developed this theory, (he is also famous for his work with electricity – Coulomb's Law – and other scientific achievements). He solved the problem of differing lateral forces for varying assumed slip planes, by use of differential calculus. His solution is the well-known Coulomb equation. (See page 25). His equation provides for varying backfill slopes, batter of the wall, and friction between the soil wedge and the face of the retaining wall.

To keep a soil wedge in place, the three forces shown in the free-body diagram in Figure 3.1 must be in equilibrium. The three forces are the weight of the wedge "W", which is its area times the soil density, and which acts vertically downward; the reaction "P" against the wall, which is assumed to have a direction inclined at the wall friction angle, and the reaction from the soil behind the wedge "F".

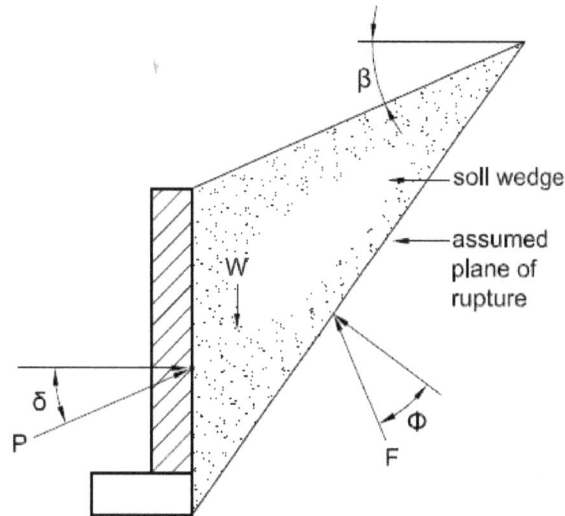

Figure 3.1 Free-body of soil wedge

Later, Scottish engineer William Rankine simplified the Coulomb equation. In the 1850s he presented the equally well known Rankine equation that neglects wall friction and is more therefore, conservative than the Coulomb method. His equation takes into account the effect of a backfill slope, assumes the back face of the stem is vertical, that there is zero friction at the soil-stem interface along the height of the wall, and the resultant acts against the wall parallel to the backfill slope.

Explanation of Design Terms:

Some commonly used terms that apply to retaining walls are defined in the following:

The rupture (or failure) plane: This is the plane along which the soil wedge is assumed to slip. It is actually concave, but a straight line is assumed to allow for the concurrency of forces; that is a necessary assumption of equilibrium. The angle this plane makes with the horizontal is theoretically: $\alpha = 45° + \phi/2$. Where α is the counter–clockwise angle from horizontal to the rupture plane. The Coulomb equation provides an estimate of the lateral force P in terms of the angle of internal friction of the soil, the back slope β and the wall friction δ.

Angle of internal friction: This is the most important value for determining lateral pressure and bearing capacity of granular and cohesive soil. It is a measure of the shearing resistance of the soil because of intergranular friction, obtained from one of several laboratory tests. Angles of internal friction range from 32-36° for well graded sand, 28-32° for silty sand, less for sandy silt, and further diminishes for clay because of the lack of coarse particles. The angle of internal friction (usually designated ϕ) is used in both the Rankine and Coulomb equations.

SOIL TYPES AND COMMON RANGE OF DESIGN VALUES		
SOIL TYPES	Phi Degrees	unit weight pcf
GRAVEL well-graded	38 to 42	120 to 140
Coarse SAND well-graded	36 to 40	105 to 135
Medium SAND well-graded	34 to 38	115 to 118
Fine SAND well-graded	32 to 36	105 to 112
Fine SAND uniform	30 to 34	95 to 110
Silty SAND	28 to 32	90 to 110
Clayey SAND	24 to 26	90 to 105

NOTES:

1. Range in values shown are for increase in phi angle with increase in density.
2. If a backfill soil does not have any cohesion, then the backfill slope β Figure 3.1, cannot exceed "phi" for a backfill of any height. If the soil has cohesion and phi values, then there is a limited height of the backfill, and that is a slope stability question for the project geotechnical engineer. However, most grading ordinances limit the backfill to 2:1, so most clayey sands or sandy clays will safely stand for reasonable heights of backfill if the contractor selects a clayey soil that can be well-compacted with readily available compaction equipment.
3. Clayey soil is not recommended for backfill.

Figure 3.2 Soil types and design values

Active soil pressure: This is the unit pressure, expressed in pounds per square foot per foot of depth, psf/ft (pcf), imposed against the wall by the wedge of soil behind the wall. It is mobilized when the wall tilts (or slides) and the wedge slides down along the rupture plane. The active lateral pressure is assumed to increase linearly with depth. Its value also increases with increasing backfill slope, because the volume of the wedge of soil increases (see Figure 3.3). The active pressure is the product of coefficient of active pressure (K_a See pages 25, 26) multiplied by the soil density. K_a for a level backfill is generally close to 0.30 for free draining backfill materials. The Geotechnical engineer generally gives this value. Multiplying a soil density of 110 pcf by a K_a value of 0.27 would, for example, result in the oft-used lateral pressure of 30 pcf for free-draining backfill materials. Also see ASCE 7-10, Table 5-1, for Design Lateral Soil Loads shown in Figure 5.4, which specifies a minimum of 35 pcf for sandy soil and up to 85 pcf for clayey soil.

The active pressure is usually provided by the geotechnical engineer as an equivalent fluid pressure (EFP), or can be computed from the Rankine or Coulomb equations if the soil angle of internal friction and, if applicable, the wall friction angle, are given and backfill slope angle is known. It is assumed to have a triangular distribution with zero at the ground surface and a maximum pressure at the bottom of the stem (for stem design) or bottom of footing (for overturning design). The pressure diagram will be trapezoidal if a surcharge exists, see page 29.

The line of action of the resultant for the Rankine equation is assumed to act parallel to the backfill slope whose angle is usually noted as β. For the Coulomb method, the force on the stem acts at an angle from the horizontal equal to the friction angle at the soil-wall interface, δ, usually assumed to be 1/2φ to 2/3φ, plus an angle equal to the batter angle of the back face of the stem, measured from the vertical.

With a sloping backfill the active pressure on the back side of the wall will increase because of the added height of the soil wedge. See Figure 3.3.

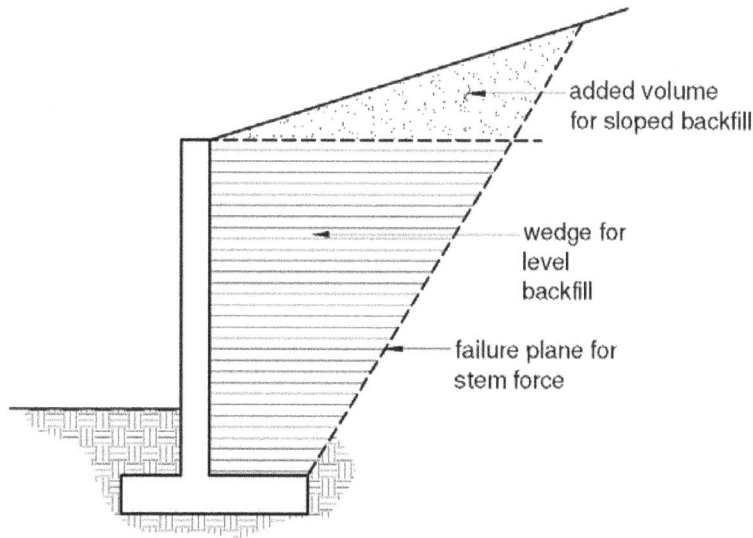

Figure 3.3 Increased wedge volume with sloped backfill

Commonly used design values (Rankine) for sloping backfills, assuming a soil density of 110 pcf, an angle of internal friction of 34° and "rounding", are:

Level:	31 pcf
5:1 Slope:	33 pcf
4:1 Slope:	34 pcf
3:1 Slope:	36 pcf
2:1 Slope:	45 pcf
1.5: 1 Slope:	77 pcf

These values are guides only and the design value should be provided by a geotechnical engineer, particularly for slopes steeper than 2:1. The slope angle cannot exceed the angle of internal friction for cohesionless soil, otherwise the slope soil will slough and not be stable.

Passive soil pressure: This is the resistance of soil to being pushed against by a rigid surface. It is obtained by multiplying the soil density by the coefficient of passive pressure, K_p. Passive pressures are usually in the range from 250 - 350 pcf. The geotechnical engineer generally gives this value. The Rankine equation for K_p is the reciprocal of K_a, ($K_p = 1 / K_a$).

Passive pressures are also assumed to vary linearly with depth.
The value of the unit weight of the backfill should reflect in-situ conditions. If the soil is dry then γ_{dry}, if wet then γ_{wet} or if saturated then γ_{sat} (γ is the common notation for soil density).

If below ground water, then $\gamma = \gamma_{sat} - 62.4$, but the designer must then include a hydrostatic lateral pressure diagram, as the total lateral force is that based on the submerged weight of the soil plus the hydrostatic force of the water.

It is good practice to limit K_p to not more than 2.5. Passive pressure is discussed further in Soil Bearing and Stability, Chapter 8, on Page 65.

"At rest" soil pressure: This lateral pressure, designated as the K_o condition, applies to non-yielding walls which are laterally supported and restrained from movement at the top and bottom, such as so-called "basement walls". This will be discussed further in the Chapter *Restrained (non-yielding) Walls*. High at-rest lateral pressures can occur if a backfill is highly compacted. Consult with the geotechnical engineer for values of K_o (See Figure 5.4).

Soil bearing values: Generally, allowable soil bearing pressures range from 1,000 psf to 4,000 psf. Additional increases are permitted for increases in width and/or depth of a footing beyond the minimum values specified by the geotechnical engineer. In general practice applicable, bearing values may be increased by one-third when wind or seismic forces are present provided that the allowable soil bearing pressures provide for a factor of safety of 1.5 or more. This increase may not be applicable to clean sands or gravels. Consult with the project geotechnical engineer on this matter. The computations of allowable bearing is part of the services provided by the project geotechnical engineer, but for those interested in the process, refer to Bowles, Foundation Analysis and Design, 5th Edition, Chapter 4, and other texts. Alternatively, subject to acceptance by the local building official, the use of presumptive values presented in the IBC '18 Table 1806.2 may be used. This table lists allowable bearing values for soil classified by the Uniform System for Classification of Soil in Appendix B.

Angle of repose: This is the slope angle, measured from the horizontal, that a stockpile of dry, granular soil will stand when loosely poured onto a flat surface; for sand, it is about 34° which is a 1.5:1 slope.

Soil density: The weight of soil is usually assumed to be 110 to 120 pcf, depending upon gradation, water content and degree of compaction. Saturated soil has a higher density, because of the added weight of water filling the voids. Soil below the water table is described as being submerged and the weight is the saturated weight minus the unit weight of water.

Backfill slope: The slope of the backfill behind and above the top of the wall cannot exceed the angle of internal friction for cohesionless soil. A general rule is to limit this slope to 2:1 (unless otherwise recommended by the geotechnical engineer).

Equivalent fluid pressure (EFP): The equivalent "hydrostatic" soil lateral pressure which increases linearly with depth. These values are the product of $K_a * \gamma$ or for passive resistance $K_p * \gamma$.

Coefficient of friction: When evaluating sliding resistance this is the frictional resistance along the contact surface between the bottom of the footing and the soil. It is a function of the roughness of the bottom of the footing, but it cannot exceed tan ϕ of the foundation soil. Its value is usually between 0.25 and 0.40, with the latter commonly used. It is used to compute resistance to sliding by multiplying the total vertical force by the coefficient of friction. This, combined with the passive pressure resistance at the toe (or toe and key), subject to a suitable factor of safety, resists sliding. Note that for cohesive soil, the resisting force is the adhesion (See page 65) between the footing and soil, rather than the frictional resistance. This cohesive force is given in pounds per square foot of contact surface, and is generally around 100 psf. Consult the project geotechnical engineer for values and applicability.

Soil modulus: Also known as the modules of subgrade reaction, designated "k", it is an indicator of the "stiffness" of a soil. It is often used to estimate the tilt of a cantilevered retaining wall. Its units are lbs. per cubic inch (lbs/in^3) and its value varies depending upon the size of the footing and the composition of the underlying soil. Load tests to determine its value are done on a one-foot square loaded plate, and the value thus obtained must be adjusted for the width of the footing in accordance with the following often used equation:

$$k = k_1 \left(\frac{B+1}{2B} \right)^2 \quad k_1 = \text{value obtained by plate test; B = footing width.}$$

The value k can vary from a low of less than 100 lbs/in^3 for loose sand to over 1000 lbs/in^3 for clay. Typical values are in the range of 150 to 300 lbs/in^3. However, when typically used in foundation design, its influence varies inversely with its 4th root so that it may not be a highly significant variable. See Table 8.7, Chapter 8 for more details and for typical values, adapted from the U.S. Army Corps of Engineers publication number TM 5-809 Structural Design Criteria for Structures Other than Buildings. Gravel and sand backfills have a value around 300 pcf. This value should be provided by the geotechnical engineer.

Tilt and Rotation of a Cantilevered Retaining Wall

A wall must either rotate or translate slightly for the soil wedge to theory to be valid, and for incipient sliding to occur. Various values have been assigned to this rotation with common values from 0.001H to 0.004H, where H is the height of the wall in inches, with the former value being generally accepted. This leads to a rule-of-thumb of about 1/8″ to 1/2″ for ten feet of wall height.

To estimate the tilt at the top of the wall one method is to compute the compression (settlement) of the soil under the toe. This requires knowing the modulus's of subgrade reaction for the soil type. Then by geometry this simplified method to estimate the tilt at the top of the wall becomes:

Tilt (inches) = [(toe soil bearing value in psf) / 144 * (soil modulus)] * H / W, where H = wall height and B = footing width. "Tilt" is the lateral out-of plumb deflection at the top of the wall.

Note that under some loading conditions the maximum soil pressure could be at the heel, thereby reversing the tilt into the retained soil and nullifying movement.

More in-depth information on rotation can be found in Bowles' 5th edition, pages 310-311 and 501-503. Also on page 66 of this work for further discussion.

Hugh's Pickle Jar Test

The Pickle Jar test is an interesting and informative way to learn about soil, and to give an approximate classification of the soil on a project site. Take a tall slender pickle jar with clear glass and a capacity of at least two cups (16 oz.), but a similar jar will do. Scoop up a sample of soil to fill the jar about half full. Pour water into the jar until the water reaches the soil surface. Now the soil is saturated so you can visualize the voids. To maintain the water level you will probably have to add a volume of water equal to about one-fourth of the volume of sample soil.

Now pour in more water, screw the lid on tight, and shake vigorously for 30 seconds to mix the soil and water. Let the jar stand for 30 minutes. Observe how the soil settles and stratifies. Clear lines of stratification will be visible. Estimate the percentage of each layer: gravel to sand on the bottom, overlain with silt and probably a thin layer of clay on top, and maybe some floating organic debris. Now classify the soil sample by comparing it with the Uniform System for Classification of Soil (see Appendix B).

Now remove the lid and push a broad-faced knife to the bottom. Wiggle it side-to-side and watch the pressure bulge. Then slowly withdraw it and notice the friction resistance. Fold a paper towel over the top and turn the jar upside down to drain the water. Watch the soil cling to the sides (adhesion). Let it dry for a few days (don't use the microwave!) then shake it up and pour it out onto a flat surface. The slope of the soil is the angle of repose.

Play some more, it's a learning experience!

The Precision Illusion

Let's not fool ourselves. Even though the science of soil mechanics is well developed and reasonably well understood, it is not an exact science and remains both an art as well as a science. Soil is a mixture of earth materials and although its engineering properties can be reasonably defined and evaluated, its actual in situ behavior likely will not precisely reflect theory. For example: the straight line assumption for the rupture plane is actually somewhat concave; and, the "equivalent fluid pressure" of soil is not truly triangular. These are simplifying assumptions made to solve a problem that is highly indeterminate. As the adage goes: engineering is an exact science based upon assumptions. Our calculations are the best we can do with assumptions we make based upon observed soil behavior and the results are never fully accurate. That's why we use factors of safety. So keep precision in mind when calculating beyond the first decimal point. Spoil behavior doesn't always follow the rules.

What Building Code(s) Apply To This Project?

Your design must conform to applicable building codes. Always check with the Building Department having jurisdiction over the project to determine the code(s) adopted and if any local amendments apply. Often the adapted code may not be the most current addition. Also note that most major cities (Los Angeles, Chicago, New York and others), are based upon on the International Building Code (IBC) with amendments for local issues.

The following codes are most often adopted or cited.

Building Codes

International Building Code (IBC)

This standard building code has been adopted by most jurisdictions, some with local modifications (*California Building Code*, for example). The IBC was a culmination of efforts to merge into one national building code the *Uniform Building Code, Southern Building Code,* and *Standard Building Code.* The IBC is compiled and published by the *International Code Council* (ICC), County Club Hills, Illinois. The series of International Building Codes (e.g. plumbing, electrical, etc.) are collectively referred to as the "I-Codes". The IBC Website is www.iccsafe.org. The current edition is 2015. IBC 2015 references or modifies other standard codes, principally ASCE 7-10 *Minimum Design Loads for Buildings and Other Structures.*

At this writing (3/2018) IBC 2018 is available as is ASCE 7-16 but their general adoptions by jurisdictions will take time.

Uniform Building Code (UBC), '97

This now defunct code, the last in a series first published in 1927 by the *International Conference of Building Officials*, was the dominant code in the Western States until replaced by the *International Building Code* and *California Building Code.*

California Building Code (CBC)

This California Code was first published in 2001 to replace the '97 *Uniform Building Code.* It is an adaptation of the IBC with minor modifications and is essentially the same as the IBC. The current edition is 2016.

NFPA 5000: Building Construction and Safety Code (National Fire Prevention Association)

NFPA 5000 has been promoted in some States. It addresses construction protection and occupancy features necessary to minimize danger to life and property. The current edition is *NFPA 5000: Building Construction and Safety Code*, 2015 Edition. The NFPA web address is www.nfpa.org. This Code references ACI 318-14, ASCE 7-10 and TMS 402/602-16 for structural design requirements.

Referenced Publications

The International Building Codes (IBC), and regional codes, often refer to the following for more specific for structural requirements:

Minimum Design Loads for Buildings and Other Structures, ASCE/SEI 7-10
Published by American *Society of Civil Engineers* (ASCE), Reston, VA. This often referenced publication covers loads and seismic design. See www.asce.org. Note that ASCE 7-16 is available but not widely adopted at present (3/2018). Its general adoption may correspond to adoption of IBC 2018.

Building Code Requirements for Reinforced Concrete (ACI 318-14), *American Concrete Institute* (ACI 318-14), Detroit, MI. Standard for concrete design. www.concrete.org

Building Code Requirements and Specifications for Masonry Structures (TMS 402/602-16)
Formerly know as ACI 318-14-530 until rights were acquired by The Masonry Society (TMS). See www.masonrysociety.org.

National Earthquake Hazard Reduction Program (NEHRP), developed by the Building Seismic Safety Council for FEMA (Federal Emergency Management Agency). This is not a code, per se, but is referenced by IBC and NFPA as guidelines for seismic design. The 2015 Edition *NEHRP, Recommended Seismic Provisions for New Buildings and Other Structures,* contains information on seismic design of retaining walls and in the Commentary. See www.nehrp.gov.

Annual Book of ASTM Standards. This is the standard of reference on materials and processes cited in most codes and specifications. Its 70+ volumes cover over 11,000 specifications. Published by *ASTM International,* See www.astm.org.

National Design Standards for Wood Construction (NDS), 2015. Published by American Wood Council www.awc.org.

Other Codes as applicable:

AASHTO LRFD Bridge and Highway Design Specifications, 7th. Edition, 2014, American Association of State Highway and Transportation Officials (AASHTO), www.aashto.org.

Naval Facilities Engineering Command (NAVFAC). *Soil Mechanics, Foundations and Earth Structures*, NAVFAC DM 7. This design manual contains information on many aspects of retaining structures. One source to download DM 7.02 is: http://www.geotechnicaldirectory.com/publications/Dm702.pdf. Many Navy documents are available for download.

U.S. Army Corps of Engineers Design Manuals. Comprehensive design procedures, Standards, and sample calculations: The web address is: www.usace.army.mil. Many Corps documents are available for download.

CRSI Design Handbook, 2012. Concrete Reinforcing Steel Handbook, www.crsi.org.

5.   FORCES AND LOADS ON RETAINING WALLS

Determination of Loads and Forces

The design of retaining walls may include any or all of the following (each will be discussed in the text that follows):

- Lateral earth pressure
- Axial loads
- Adjacent footing loads
- Surcharge loads
- Impact forces
- Wind on projecting stem
- *Seismic wall self-weight forces and seismic earth pressure force
  *Discussed in Chapter 6

Lateral Earth Pressures

The purpose of a retaining wall is to retain soil and to resist the lateral pressure of the soil against the wall. Most lateral pressure theories are based upon the sliding soil wedge theory. This, in simple terms, is based upon the assumption that if the wall is suddenly removed, a triangular wedge of soil will slide down along a rupture plane, and it is this wedge of soil that the wall must retain. There are two basic equations for computing lateral earth pressures: the Coulomb equation and the Rankine equation.

The Coulomb Equation

This classic equation was developed by Charles Coulomb (17 30 – 1806) well known for other scientific advancements. The Coulomb Equation, where K_a is the coefficient of active pressure, takes into account the backfill slope, friction angle at wall face, angle of rupture plane and angle of internal friction. See Figure 5.1:

$$K_a = \frac{\sin^2(\alpha + \phi)}{\sin^2\alpha \, \sin(\alpha - \delta)\left[1 + \sqrt{\frac{\sin(\phi + \delta)\sin(\phi - \beta)}{\sin(\alpha - \delta)\sin(\alpha + \beta)}}\right]^2}$$

$$K_a \text{ (horiz.)} = K_a \cos\delta \ \text{ if } \alpha = 90°$$

β = Angle of backfill slope
ϕ = Angle of internal friction of the soil
α = Wall slope angle from horizontal (90° for vertical face, α = < 90° if the back of wall is battered outward or > 90° if wall battered inward)
δ = Angle of friction between soil and wall (usually assumed to be 2/3ϕ to 1/2/ϕ)

Figure 5.1 The Coulomb equation

The Coulomb equation should only be used for gravity, segmental, gabion, and cantilevered walls having a short heel dimension. The reason is that the Coulomb equation includes a soil-to-wall friction angle , designated δ, which assumes the moving soil mass contacts the wall face and activates a shear resistance as the wall deflects. This friction angle δ is generally assumed to be between 0.5 and 0.7 times the phi (φ) angle. For the case of a cantilevered wall with a larger heel dimension the soil between the stem and the failure plane can be considered a rigid mass, then δ in the Coulomb equation or Mononobe-Okabe equation (discussed in the seismic chapter) can be taken a equal to φ because with a cantilevered wall the soil above the heel will move in mass with the wall so that wall friction cannot develop, the failure plane being through the heel of the footing.

If the backfill is level, the inside wall face is vertical, and if zero friction is assumed between the soil and wall, then the Coulomb equation reduces to the familiar Rankine equation:

$$K_a = (1 - \sin \phi) / (1 + \sin \phi)$$

The Rankine Equation

William Rankine (1820 – 1872) developed the well known Rankine equation is a simplified version of the Coulomb equation that does not take into account wall batter or friction at the wall-soil interface. As such, it is a conservative approach to the design of retaining walls. An example of its use will be described later for both the Coulomb and Rankine equations. Vertical walls with a level backfill and zero wall friction, the lateral pressure factor K_a will be the same by either approach

Rankine's approached the evaluation of the stress at a point in a backfill by using Mohr's circle concepts to obtain the minimum lateral stress at a point in the backfill. The minimum lateral stress corresponds to the "active" case. Integration of that stress with respect to depth leads to a second-order equation (the well-known triangular distribution) for the total lateral force against the wall.

The use of the Rankine approach is recommended for most cantilevered retaining wall designs. It is conservative because it predicts a larger active force than that of Coulomb. It's also simpler to calculate for most walls, and easily handles sloping backfills and surcharge loads.

The Rankine Equation for active pressure:

$$K_a = \cos \beta \; \frac{\cos \beta - \sqrt{\cos^2 \beta - \cos^2 \phi}}{\cos \beta + \sqrt{\cos^2 \beta - \cos^2 \phi}}$$

$$K_a = (\text{horiz.}) = K_a \cos \beta$$

β = Angle of backfill slope
φ = Angle of internal friction of the backfill soil

Figure 5.2 The Rankine equation

If the backfill is level the Rankine equation can be written as: $K_a = \tan^2\left(45 - \dfrac{\phi}{2}\right)$ or $= \dfrac{1 - \sin\phi}{1 + \sin\phi}$

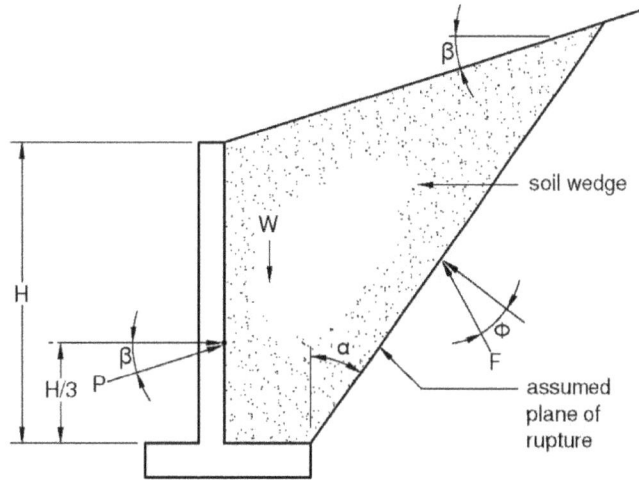

Figure 5.3 Rankine free-body of lateral forces on stem

Example: Assume: $\phi = 34°$, $\beta = 26.6°$ (2:1 slope)

Then $K_a = \dfrac{\cos 26.6 - \sqrt{\cos^2 26.6 - \cos^2 34}}{\cos 26.6 + \sqrt{\cos^2 26.6 - \cos^2 34}} \cos 26.6$

$= 0.41$

If the backfill is sloped convert K_a to its horizontal component for computing stem moments and overturning.

Therefore K_a horiz. $= 0.41 \times \cos 26.6 = 0.37$ and corresponding horizontal equivalent fluid weight of the soil $= 0.37 \times$ say 110 pcf $= 40$ pcf for a 2:1 backfill slope ($= 26.6°$).

Code Prescribed Lateral Pressure

IBC '18 Table 1610.1 and ASCE 7-10 have similar tables of minimum lateral pressures for level backfill, condensed in Figure 5.4. Note that ASCE 7-16, Table 3.2-1 shows somewhat higher active pressure for some soils shown in parenthesis.

Backfill Material	USCS Classification	Lateral Pressures (pound per square foot per foot of depth)	
		Active pressure	At-rest pressure
Well-graded, clean gravels; gravel-sand mixes	GW	30 (35)	60
Poorly graded clean gravels; gravel-sand mixes	GP	30 (35)	60
Silty gravels, poorly graded gravel-sand mixes	GM	40 (35)	60
Clayey gravels, poorly graded gravel-and-clay mixes	GC	45	60
Well-graded, clean sand gravelly sand mixes	SW	30 (35)	60
Poorly graded clean sand sand-gravel mixes	SP	30 (35)	60
Silty sands, poorly graded sand-silt mixes	SM	45	60
Sand-silt clay mix with plastic fines	SM-SC	45 (85)	100
Clayey sands, poorly graded sand-clay mixes	SC	60 (85)	100
Inorganic silts and clayey silts	ML	45 (85)	100
Mixture of inorganic silt and clay	ML-CL	60 (85)	100
Inorganic clays of low to medium plasticity	CL	60 (100)	100

Figure 5.4 Lateral soil pressures (Condensed from IBC 2018 and ASCE 7-16)

Surcharge Loads

A surcharge is any additional vertical load applied to the backfill soil. It can be a live load from a parking lot or highway, paving or an adjacent footing. See Figure 5.5 to illustrate this affect on lateral pressures.

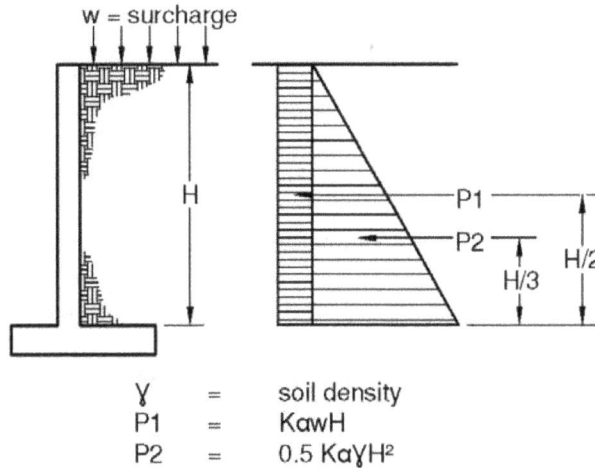

$$
\begin{aligned}
\gamma &= \text{soil density} \\
P1 &= Ka w H \\
P2 &= 0.5\, Ka \gamma H^2
\end{aligned}
$$

Figure 5.5 Active pressure on the stem from a uniform surcharge above the wall

Lateral Pressure from Mixed Soil

Backfill could be composed of several layers of soil with differing characteristics. A method for computing the lateral pressures from each layer is shown in Figure 5.6. Summing the lateral pressure from each layer gives the total at the base of the stem. In this Figure the overturning moment at the base would be $[P_1 * H_1 + P_2 * H_2 + P_3 * H_3 + P_4 * H_4 + P_5 * H_5]$, where P_x are the various lateral force components and H_x is the height to the centroid of each. Similarly, moments and shears at any height for stem design can be computed.

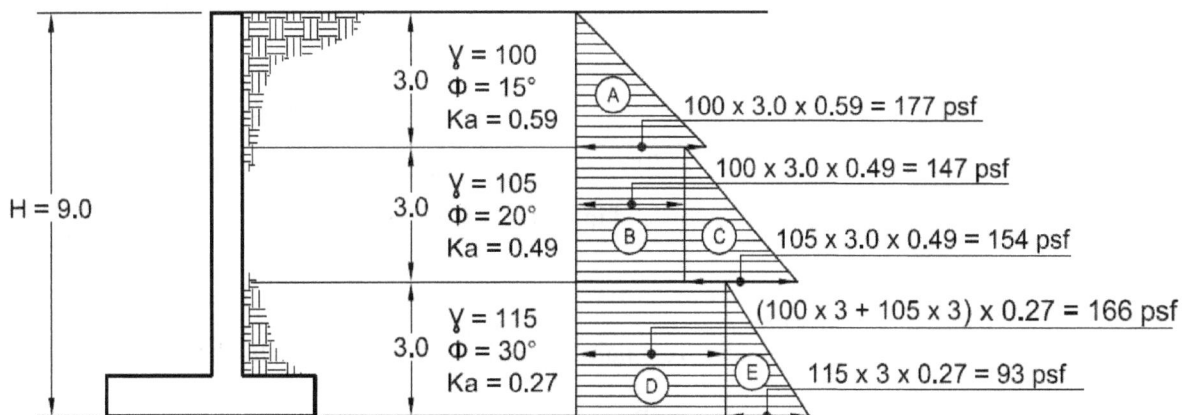

Figure 5.6 Lateral soil pressure from mixed soil type layers

Highway surcharges

The usual added surcharge when highway traffic is close to a retaining wall (within less than one-half the height from top of footing to applied surcharge) is 250 psf. This is an AASHTO requirement.

Backfill compaction surcharge

Backfill is often placed by a front end loader dumping sand or gravel behind the wall. The backfill should be placed in layers of about one foot thick, compacted by repeated by back-and-forth runs of a compactor or loader, coming within inches of the wall. Compaction testing may be required. A typical loader weighs about 30,000 lbs, and has a footprint under each track of about 30 square feet if the loader is track-mounted. This results in short-time construction loads of about 1000 psf, far in excess of most surcharge design loads. Grading contractors are aware of this and often report tilting of the wall during these operations, and sometimes assign a worker to monitor plumb of the wall during these operations. Backfill compaction can produce high lateral pressures or a K_o condition ("at-rest pressure") if the wall is restrained. It should be the contractor's responsibility to place backfill so as to not damage or overstress a wall and if in doubt consult the structural engineer for a design review.

Adjacent footing surcharges

If there is an adjacent footing that overlays the area of the soil wedge this surcharge will exert lateral pressure against the wall, which must be considered. A rule-of-thumb is that an adjacent footing will have little effect on lateral pressure against the stem if it is further than the height (base of stem to base of applied adjacent load) away from the wall face. Adjacent footing loads are classified as either "line loads" of unit width, "strip" loads" which are uniform loads parallel to the wall as from a wall footing, or "point loads" such as square or rectangular footings. Solutions for these cases are presented in the following paragraphs.

Boussinesq Method

The Boussinesq equation, developed by Joseph Boussinesq in 1885 and derived from the theory of elasticity, is computationally very laborious but is often used to evaluate the influence of adjacent loads on a wall as shown in Figure 5.7.

The Boussinesq equation:

$$\sigma_r = \frac{P}{2\pi}\left(\frac{3r^2 z}{R^5} - \frac{1 - 2\mu}{R(R + z)}\right)$$

Where terms are defined below and in Figure 5.7 for a point load:

σ_r = Lateral Pressure, psf
P = Point load, lbs.
r = Horizontal distance from point of application on wall to plumb under P
z = Depth, ft.
R = Diagonal distance from P to point of application on the wall.
$= \sqrt{r^2 + z^2}$
μ = Poisson's ratio

Also, note that the Boussinesq distribution has the shape shown in Figure 5.8 is sensitive to the assumed Poisson's Ratio (μ) for the soil. The value for sand and sandy-clay ranges from about 0.2 to 0.5. The Bowles text is an excellent reference on the use of the Boussinesq equations. Carefully note that the equation does not contain a soil modulus term, such as the modulus of elasticity.

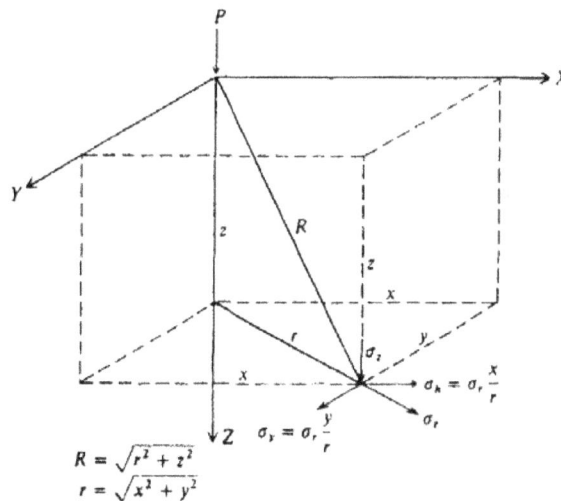

$$R = \sqrt{r^2 + z^2}$$
$$r = \sqrt{x^2 + y^2}$$

Figure 5.7 Terms for Boussinesq Equation

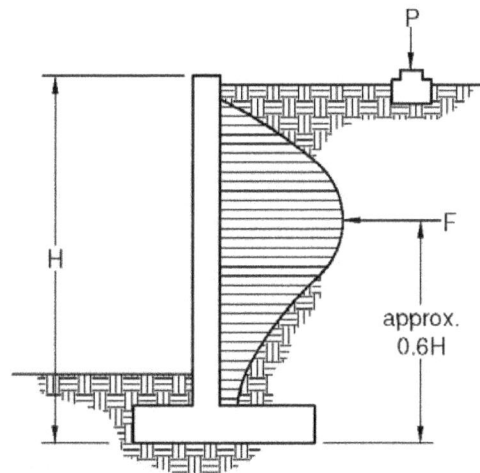

Figure 5.8 Boussinesq equation showing lateral pressure curve

NAVFAC Method

NAVFAC DM-7 presents often used equations for lateral pressure from point and line loads and includes charts to make the task easier. These equations are illustrated in Figure 5.9. These equations, also available from other texts, are from the Boussinesq equation modified by experiment.

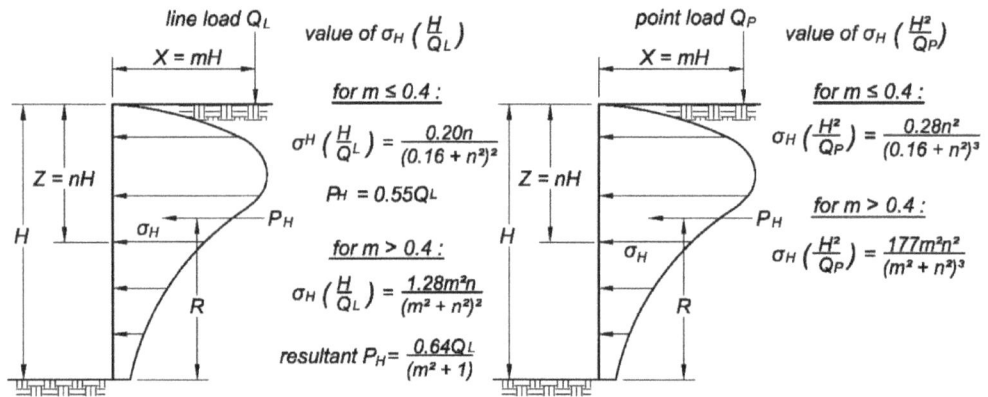

Figure 5.9 NAVFAC pressures from line and point loads

Vertical Surcharge from Line Load

Terzaghi and Peck in *Soil Mechanics in Engineering Practice*, on page 367, propose a simplified method for computing the vertical stabilizing force on the footing to include only the surcharge directly above. See Figure 5.10.

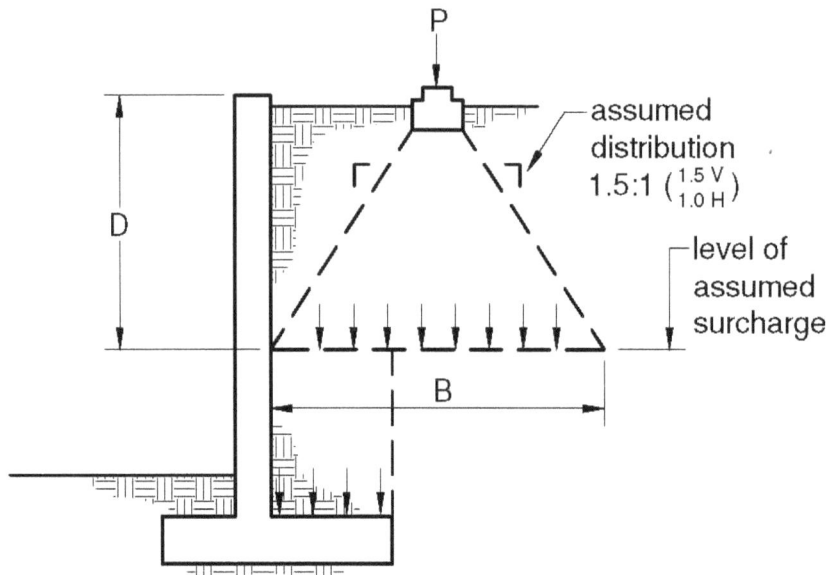

Figure 5.10 Vertical pressure from line load

The above discussion of adjacent footing loads is brief; more in-depth treatments can be found in the texts referenced in the bibliography, Appendix H.

It must be pointed out that there is a wide divergence of values from the various theoretical lateral pressure methods versus in-situ tests that vary in reliability because of the procedures used. Some even suggest that results from Boussinesq be doubled. Good judgment should be your guide.

Wind Load on Free-standing Walls and Exposed Projecting Stems

Design for wind forces can be complex as designers have discovered from IBC '18, 1609, and ASCE 7-16, section 26-31. Inasmuch as an exemption is made for wind forces on fences not over seven feet high (IBC '18, Sections 105.2 and 312.1), arguably an exception may apply to projecting portions of a wall above its retained height. Such walls could be designed for the minimum wind force of 16 psf (IBC '18, 1609.6.3). However, the designer must use judgment considering interpretation of applicable code provisions, site topography, and local wind information available.

Water Table Conditions

If a portion of the retained height is below a water table, the active pressure of the saturated soil will increase below that level. This additional pressure for the saturated soil is equal to the pressure of water, plus the submerged weight of the soil (its saturated weight - 62.4), plus the surcharge of the soil above the water table. The submerged weight of a soil can be approximated as 62% of its dry unit weight. This pressure diagram is shown in Figure 5.13.

Assume: $K_a = 0.27$
$\gamma_w = 95$ pcf
γ saturated $= 110$ pcf
Surcharge - 100 psf
Water $= 62.4$ pcf

Lateral Pressures:

0.27×100 psf $= 27$ psf.
$0.27 \times 95 \times 6' = 154$ psf
$0.27(110-62.4) \times 7.5' = 96.4$ psf
$62.4 \times 7.5' = 468$ psf

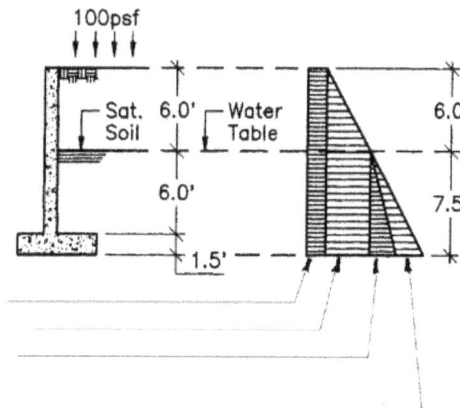

Summary for overturning:

P			$\overline{X}$	M
27×13.5	$=$	$365^{\#}$	$6.75'$	$2,464^{'\#}$
$154 \times 6/2$	$=$	$462^{\#}$	$9.5'$	$4,389^{'\#}$
154×7.5	$=$	$1155^{\#}$	$3.75'$	$4,331^{'\#}$
$96.4 \times 7.5/2$	$=$	$362^{\#}$	$2.5'$	$904^{'\#}$
$468 \times 7.5/2$	$=$	$1755^{\#}$	$2.5'$	$4,388^{'\#}$
		$4099^{\#}$		$16,476^{\#}$

Height to point of application of the resultant horizontal force = 16,472 / 4,099 = 4.02

Figure 5.11 Water table force calculations

Detention Ponds and Flood Walls

For retaining liquids, the procedure is similar to an earth backfill except the equivalent fluid pressure is 62.4 pcf (or that of the liquid) and weight on the heel is the same fluid weight for overturning and soil pressure calculation.

Hydrostatic Pressures

If the water table is above the foundation on the heel side as in Figure 5.11 and seepage under the footing is likely, then there will be uplift on the underside of the footing. The variation in the uplift pressure across the bottom of a footing must be evaluated, usually by a conservative approach assuming a uniform distribution. Consult with the project geotechnical engineer on this. When performing stability analysis the footing weight may have to reduce to account for buoyancy. Also, the allowable soil bearing pressure, which is a function of soil density, will thereby be less than those indicated in the geotechnical reports, is unless addressed in the report.

Cascading Walls

Occasionally walls will be stacked one behind another, piggyback style, or cascading, as sketched in Figure 5.12. This situation requires very careful design for the lower walls because not only is there a surcharge from the wall above but a horizontal thrust as well. Two possible solutions are suggested:

Alternate #1 shown on Figure 5.12, would be to sketch a fail-safe slope that would model the event if both upper walls were considered one mass exerting pressure on the lower wall.

Another solution often used, Alternate #2, is to apply the vertical load P_v from an upper wall as a line-load surcharge on a next-lower wall, using the Boussinesq equation, as force P_v located "x" from the lower wall, then apply its horizontal thrust P_h as an assumed uniform load against the stem of the lower wall.

Cascading wall conditions come up frequently in practice and a good reference for that design is not known to the authors. Carefully consider the horizontal thrust from higher walls.

For cascading walls the geotechnical engineer must perform a global stability analysis (see Chapter 8 page 67) depending upon the nature of the underlying soil.

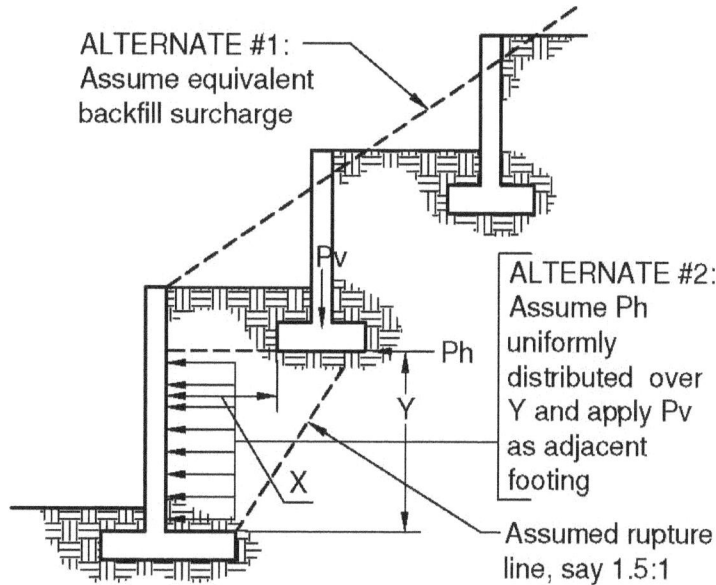

if X > Y probably little effect on lower wall

Figure 5.12 Cascading walls

Vertical Loads

These loads affect design for stability and components of the wall. Vertical loads include:

Axial dead loads on the stem: These loads are applied directly to the stem, such as from a beam reaction, ledger, or bridge member. Any vertical load imposed upon the stem of a cantilever retaining wall must not provide lateral support to the wall, otherwise the wall does not perform as a cantilevered retaining wall, that is, the lateral soil pressures will not reduce to the K_a condition. If one side of a building, for example, rests on top of a wall—it could be reactions from a floor or roof—the abutting floor diaphragm especially those of wood construction may not restrain lateral movement (rotation) of the wall. If restraint does occur, the wall should be designed as a "basement wall" (K_o condition), whereby the restraint at the top results in a positive bending moment near mid-height the stem. Sometimes a wall is designed for both conditions, such as when it is designed as a retaining wall so that backfill can be placed before the restraint is provided (See Backfill Compaction Surcharge, page 30), then designed as a basement wall for the condition after the restraint is in place.

Axial live loads on the stem will increase the soil bearing pressure and resisting moments, therefore live loads need to be assessed separately from axial dead loads to determine the most critical condition.

Vertical point loads on walls, such as girder reactions, are assumed to spread downward along, the longitudinal length of the wall at a slope of two vertical one horizontal. This spreading of the load results in relatively low compressive stresses at the base of the stem. For example, a 24 kip load atop a 12" concrete wall on a two-foot wide bearing, and assuming 14 ft high, would result in an axial compressive stress (in addition to wall weight) of just 125 lbs/per linear inch of wall at the base of the stem. Bearing stresses directly at the top of the stem under a beam or girder reaction must also be checked.

Also, consider the eccentricity of the girder reaction with respect to centerline of the stem because it will affect both stem design and stability. Consider, however, that live loads acting at a negative eccentricity (toward the backfill) could produce unconservative results with respect to the movement stability of the wall.

Axial loads rarely exceed about 2,000 lbs/ft. but usually are much less. Caution must be used if for some reason a very high axial load is applied because it could change the bending characteristic of the footing. For example a very high heel soil pressure because of a high axial load could reverse the typical bending stress in the heel requiring significant reinforcing at the bottom (tension) rather than the top.

Weight of soil: This includes the weight of the soil over both the heel and toe.

Weight of structure: This includes the weight of the stem and footing and that of the key.

Axial and surcharge loads: These loads add to soil bearing pressure and may contribute to overturning and sliding stability.

Vertical component of active pressure: If there is a sloped backfill the lateral forces will have both a vertical and horizontal component. The former is assumed to act on a vertical plane at the back of the heel. See page 61. However, the stabilizing moment of this force is usually neglected in design because of the uncertainties associated with a uniform strain occurring along the assumed failure plane.

Lateral Impact Loading

If the wall extends above grade with an adjacent parking area, the designer may want to design for impact from a vehicle. ASCE 16, 4.5.3 specifies 6,000 lbs. applied at a height of 18″ above grade. Per code, guard rail design requires 50 plf applied to the top, or a single concentrated lateral load of 200 lbs. Verify requirements with applicable codes.

When considering the lateral effect of impact the stem should be checked at incremental descending points because the lateral impact force spreads over a greater stem length. Assume the impact load spreads out laterally at one horizontal to two vertical. This is equivalent to spreading its effect over a length of wall equal to the distance from point of application down to the plane being checked.

Earthquakes – An Overview

Although our planet Earth is an incomprehensible 4.5 billion years old it is still cooling and adjusting. The tectonic plates (tectonic from the Latin: "building") that wrap our earth continue to float, move, and rotate, as in past eons, to shape our topography, build mountains, and cause earthquakes along plate boundaries.

Earthquakes can occur anywhere. However, in the United States the West Coast is most vulnerable as the Pacific tectonic plate, which covers the entire Pacific Rim, rotates counter-clockwise, northward along the West Coast, moving about an inch per year as it grinds past its boundary with the easterly North American plate. This movement is primarily along California's infamous San Andreas Fault (so named for the community it passes near San Francisco) and is responsible for the numerous stress-relieving earthquake jolts occurring daily on the many associated faults.

In California there are over 400 measurable earthquakes each week   Many are never felt (those under Magnitude 3.0 are rarely felt), and fortunately, few cause damage. Some of the larger earthquakes in California with magnitudes greater than 5 include:

> 1908, San Francisco, 7.2 (estimated)
> 1933, Long Beach, 6.4 (estimated)
> 1940, Imperial Valley, 7.0
> 1952, Kern County, 7.3
> 1971, San Fernando, 6.7
> 1987, Whittier Narrows, 5.9
> 1989, Loma Prieta, 6.9
> 1992, Landers, 7.3
> 1994, Northridge, 6.7
> 2010, Borrego Springs, 5.4

> 20xx, The California "Big One": who knows when or where?

> Note: Reports of earthquakes prior to 1935 use estimated Richter magnitudes.

Ironically, however, one of the largest earthquake events occurred mid-continent, near the town of Madrid on the Mississippi River midway between St. Louis and Memphis. Known as the New Madrid faults, there were a series of earthquakes in 1811-12 with estimated magnitudes of ≈ 7.7. They were felt as far away as New York City and reportedly rang church bells as far away as Boston.

The largest recorded in North America: 1964, Alaska, 9.2

The largest earthquake ever recorded worldwide was in Chile in 1960 with a magnitude of 9.5.

More recently on 3/11/2011 off-shore Japan, magnitude 8.9. On 9/19/2017 southern Mexico, magnitude 7.1.

The term "magnitude" as used in the above list, and in the media, was developed in 1935 by Caltech professor Charles Richter and colleagues and bears his name "Richter Scale". They used data from a seismograph to describe a specific earthquake in terms of seismic energy released. It is a logarithmic scale (to the base 10) whereby a magnitude 5 earthquake releases about ten times the energy of that of a magnitude 4 ($5^{10} / 4^{10} \approx 10$). This measure is popular with the media but does not have a direct correlation to ground acceleration that is used to determine the seismic force for the design of structures.

Prior to using the Richter Scale, the Mercalli Intensity Scale was developed, which classified earthquakes based upon their effect at the earth's surface. It was developed by Guiseppe Mercalli in 1902 and described in a USGS pamphlet as shown below – the higher the number the more severe the damage:

(I) - Not felt except by a very few under especially favorable conditions.

(II) - Felt only by a few persons at rest, especially on upper floors of buildings. Delicately suspended objects may swing.

(III) - Felt quite noticeably by persons indoors, especially on the upper floors of buildings. Many do not recognize it as an earthquake. Standing motor cars may rock slightly. Vibration similar to the passing of a truck. Duration estimated.

(IV) - Felt indoors by many, outdoors by few during the day. At night, some awakened. Dishes, windows, doors disturbed; walls make cracking sound. Sensation like heavy truck striking building. Standing motor cars rocked noticeably.

(V) - Felt by nearly everyone; many awakened. Some dishes and windows broken. Unstable objects overturned. Clocks may stop.

(VI) - Felt by all; many frightened and run outdoors, walk unsteadily. Windows, dishes, glassware broken... books off shelves... some heavy furniture moved or overturned; a few instances of fallen plaster. Damage slight.

(VII) - Difficult to stand... furniture broken… damage negligible in buildings of good design and construction; slight to moderate in well-built ordinary structures; considerable damage in poorly built or badly designed structures; some chimneys broken. Noticed by persons driving motor cars.

(VIII) - Damage slight in specially designed structures; considerable in ordinary substantial buildings with partial collapse. Damage great in poorly built structures. Fall of chimneys, factory stacks, columns, monuments, walls. Heavy furniture moved.

(IX) - General panic... damage considerable in specially designed structures, well designed frame structures thrown out of plumb. Damage great in substantial buildings, with partial collapse. Buildings shifted off foundations.

(X) - Some well-built wooden structures destroyed; most masonry and framed structures destroyed with foundation. Rails bent.

(XI) - Few, if any masonry structures remain standing. Bridges destroyed. Rails bent greatly.

(XII) - Damage total. Lines of sight and level distorted. Objects thrown into the air.

The Richter magnitude scale, used mostly by the media and for general intensity comparisons, is now replaced by site-specific ground accelerations as explained in following sections. An excellent source of information on earthquakes, including hazard maps, is http://www.usgs.gov.

How have Earthquakes affected Retaining Walls?

Not much, according to numerous post-earthquake investigations by the author and other engineers, including San Fernando 1971, Loma Prieta 1989, and Northridge 1994. Predictably, damage was observed in walls improperly designed, site soil problems, liquifaction and other problems.

Reports of damage to basement (restrained) walls are virtually nil.

A reason retaining walls resist seismic forces is most likely the 1.5 safety factor used for static design which could absorb all but the most powerful jolts.

However, in spite of the evidence of the seismic sustainability, the Building Codes may require design for seismic forces.

When is Seismic Design Required for Retaining Walls?

IBC 2018, Section 1807.2.2 states: . . . *For structures assigned to Seismic Design Category D, E, or F, the design of retaining wall supporting more than 6 feet of backfill height shall incorporate the additional seismic earth pressure in accordance with the geological investigation where required by Section 1803.2.* Note: For Seismic Design Category see IBC '18, Table 1613.2.5(2).

For seismic requirements of IBC 2018, Section 1613.1, refers to ASCE 7-16 which in Section 15.6.1 *Earth-Retaining Structures* reads as follows:

ASCE 7-16 chapter 15.6.1, *Earth-Retaining Structures*: *"This section applies to all earth-retaining structures assigned to Seismic Design Category D, E, or F* (Note: these preclude Seismic Design Categories A, B, and C which are exempt from seismic design because S_{DS} is less than 0.50 -- see following section for definition of S_{DS}) *The lateral pressure due to earthquake ground motion shall be determined in accordance with Chapter 11.8.3".* This latter Chapter states that if a geotechnical investigation report is required (often at the discretion of the building official) the report shall include *"The determination of dynamic seismic lateral earth pressures on basement and retaining walls caused by design earthquake ground motions".* 15.6.1 continues:

"The risk category shall be determined by the proximity of the earth-retaining structure to other buildings and structures. If failure of the earth-retaining structure would affect the adjacent building or structure, the risk category shall not be less than that of the adjacent building or structure. Earth-retaining walls are permitted to be designed for seismic loads as either yielding or non-yielding walls. Cantilevered reinforced concrete or masonry retaining walls shall be assumed to be designed as simple flexural wall elements."

Risk Category is obtained from IBC '18, Table 1604.5. Retaining walls are generally low risk due to distance from a structure that would be damaged if the wall failed. If nearby another structure the Risk Factor would be that of the affected structure.

If a geotechnical investigation is required per IBC '18, 1803, or the local Building Official, such a report shall comply with ASCE 7-16, Section 11.8.3. requirements below.

ASCE 7-16, Section 11.8.3 states:

"The geotechnical investigation report for a structure assigned to Seismic Design Category D, E, or F shall include all of the following, as applicable:

1.   The determination of dynamic seismic lateral earth pressures on basement and retaining walls caused by due to design earthquake ground motions.

2. *The potential for liquefaction and soil strength loss evaluation for site peak ground accelerations, earthquake magnitude, and source characteristics consistent with the MCEG, peak ground acceleration. Peak ground acceleration shall be determined based on either (1) a site specific study taking into account soil amplification effects as specified to Section 11.4.7 or (2) the peak ground acceleration PGA_M from Eq. 11.8-1.*

$$PGA_M = F_{PGA} PGA \qquad\qquad (Eq. \ 11.8\text{-}1)$$

PGA_M = MCE_G peak ground acceleration adjusted for Site Class effects.
PGA = Mapped MCE_G peak ground acceleration shown in Figs. 22-7 through 22-11.
F_{PGA} = Site coefficient from Table 11.8-1.

3. *Assessment of potential consequences of liquefaction and soil strength loss, including, but not limited to, estimation of total and differential settlement, lateral soil movement, lateral soil loads on foundations, reduction in foundation soil-bearing capacity and lateral soil reaction, soil downdrag and reduction in axial and lateral soil reaction for pile foundations, increases in soil lateral pressures on retaining walls, and flotation of buried structures.*

4. *Discussion of mitigation measures such as, but not limited to, selection of appropriate foundation type and depths, selection of appropriate structural systems, to accommodate anticipated displacements and forces, ground stabilization, or any combination of these measures and how they shall be considered in the design of the structure.*

Bottom line: Check with the local building authority for applicable codes and your interpretation, and with the geotechnical engineer, if engaged, for their recommendations for your project.

Seismic Design Background

Determining with some rationale how seismic forces act on retaining walls is complex and impeded by diverse opinions, differing theoretical assumptions, and in-situ tests that don't match theoretical approaches. Researchers acknowledge the complexity of this task as code-writers try to mandate minimum design guidelines for public safety.

This effort is difficult for two reasons. As stated earlier, unlike buildings where we can learn from failures, reports of damage to reasonably well designed retaining walls (that were not designed, considering seismic forces) are nearly non-existent (waterfront walls and liquefaction conditions excluded) therefore there is little to observe and analyze to suggest design remedies. And as opined above, many question the need for adding seismic forces to static-designed retaining walls, considering both performance history and factors of safety incorporated into the design. Secondly, and compounding the dilemma, as stated above many of the theoretical approaches to determine seismic forces on retaining walls rely upon differing assumptions that yield differing results, and to in-situ and laboratory tests that didn't perform as theory predicted.

In past years "pseudo-static" (that is, using a static force to simulate a dynamic force) analyses were conducted for which the inertial effects of ground shaking were represented by a lateral *force*, which then made the problem solvable using statics. In the Uniform Building Code of 1949, for example, that force was usually set equal to 0.15W, where 0.15 was assumed to be the effective horizontal ground acceleration and W the "rigid body mass" portion of the backfill. The line of action of the force was assumed to act through the center of gravity the rigid soil mass. Factors greater than 0.15 might have been used based upon a consideration of the "importance" of the wall (now codified as the *Importance Factor*).

The practice of assuming a static equivalent horizontal ground acceleration factor is now replaced with a "site acceleration" based upon site specific spectral analyses. The concept of spectral analysis, whereby the design acceleration is based upon the characteristics and period of a structure, was introduced in the 30's and codified in the 40's. Accelerations for retaining walls, which are generally considered "short period" structures (less than 0.2 seconds), use a design acceleration given for 5% damping with a 2% chance of being exceeded in 50 years. This site-specific Peak Ground Acceleration (PGA) derived from a Maximum Considered Earthquake (MCE$_M$), is given by the geotechnical consultant or can be obtained by maps in IBC '15 or ASCE 7-10 (they are the same).

The pseudo-static approach is useful when analyzing a wall for stability – overturning, soil bearing, and sliding – but does not give the distribution of seismic lateral force incrementally on the stem. Resolving this deficiency is discussed in following sections.

Seismic Design Procedure

First step for seismic design is to determine the Peak Ground Acceleration (PGA) at the site location. This can be done by inputting the longitude and latitude of the site location, or the site address, into this USGS link: https://earthquake.usgs.gov/designmaps/beta/us. (The longitude/latitude can also be found at the jobsite by the GPS feature on your iPhone or other device.)

TIP: A popular source for hazard design criteria relative to specific site location is a subscription-based ($60 annually) website for parameters for seismic, wind and others, visit http://asce7hazardtool.online.

To calculate seismic forces on earth retaining structures, the most widely accepted and frequently cited method is the Mononobe-Okabe equation (M-O). It is an adaptation of the Coulomb equation to account for seismic forces. The Coulomb equation (discussed in Chapter 5, *Forces and Loads on Retaining Walls*) can be used to determine the lateral force on a retaining wall from earth pressure but does not include the inertial force that the soil backfill impacts on a retaining wall during an earthquake. The M-O equations take into account both horizontal and vertical ground accelerations, and provide seismic coefficients for active and passive pressure K_{AE} and K_{PE} respectively. The development of this work is based on the original work by Japanese professors Mononobe and Okabe in 1926-29.

This adaptation of the Coulomb equation to calculate the total (seismic and static) pressure, includes the variable θ, which is defined as the angle whose tangent is the ground acceleration $(\theta = \tan^{-1}\left[\dfrac{k_h}{1-k_v}\right]$), Note that if the acceleration variable θ is excluded from the M-O equation it reverts to the familiar Coulomb equation.

The Mononobe-Okabe (M-O) equations

The M-O methology provides lateral earth pressures, both active and passive for gravity and seismic forces. The assumptions underlying the M-O equations are that the soil behind the retaining wall and contained between the wall and the failure (slip) plane acts as a rigid mass; the backfill is cohesionless where a ϕ = angle of effective internal friction. The M-O equation is given in several formats, all yielding equivalent results. Presented below is from *Bowles, 5th Edition*, page 643:

K_{AE} = active earth pressure coefficient, static + seismic

$$K_{AE} = \frac{\sin^2(\alpha + \theta - \phi)}{\cos\theta \sin^2\alpha \sin(\alpha + \theta + \delta)\left[1 + \sqrt{\dfrac{\sin(\phi + \delta)\sin(\phi - \theta - \beta)}{\sin(\alpha + \delta + \theta)\sin(\alpha - \beta)}}\right]^2}$$

Where $\theta = \tan^{-1}\left[\dfrac{k_h}{1 - k_v}\right]$ θ = wall slope from (90° for a vertical face),

ϕ = angle of internal friction,

β = angle of backfill slope,

δ = wall friction angle.

α = angle at inner wall measured clock-wise from horizontal (90° for vertical inner wall)

The vertical acceleration, k_v, is generally ignored; it is conservative to do so.

The horizontal component is $K_{AE} \cos \delta$ is for a wall without a batter and without a sloping backfill. See Figure 6.3.

TIP: To simplify calculating K_{AE}, RetainPro 10 has a quick solver in this Σ feature on their tool bar.

The passive earth pressure coefficient, K_{PE} is:

$$K_{PE} = \frac{\sin^2(\alpha - \theta + \phi)}{\cos\theta \sin^2\alpha \sin(\alpha - \theta - \delta)\left[1 - \sqrt{\dfrac{\sin(\phi + \delta)\sin(\phi - \theta - \beta)}{\sin(\alpha - \delta - \theta)\sin(\alpha - \beta)}}\right]^2}$$

Note: The passive pressure coefficient decreases as seismic acceleration increases.

Figure 6.1 Mononobe-Okabe equations

Determing horizontal seismic acceleration factor, k_h

You can determine k_h by this example procedure using the USGS Hazard Maps:

For example, from charts for an assumed site location, $S_s = 1.70$

Then $S_{MS} = F_a S_s$. See IBC '15, 1613.3-3.

$F_a = 1.0$ (This is a function of soil characteristics and a function of S_s. See Table 1613.5.3(1) in IBC '15).

∴ $S_{MS} = 1.0 \times 1.70 = 1.70$ $S_{DS} = 2/3\ S_{MS} = 0.667 \times 1.70 = 1.13$

S_{DS} = Design spectral response acceleration for short period (retaining walls are considered short period).

Per ASCE 7-10, Section 1613.3.4:

$k_h = \dfrac{S_{SD}}{2.5} = 0.40 \times 1.13 = 0.45$

k_h is explained in references such as the *National Earthquake Hazard Reduction Program* (NEHRP), Part 2, Commentary, 7.5.1. *It is recommended that k_h be taken equal to the site peak ground acceleration that is consistent with design earthquake ground motions as determined in Provisions Sec. 7.5.2 (that is, $k_h = S_{DS}/2.5$. Eq. C7.5-3a and C7.5-3b generally are referred to as the simplified M-O formulation.*

To check a calculated value of K_{AE}, *Whitman (1990)* proposed this simplified method to approximate K_{AE}: $K_{AE} = K_A + 0.75\,k_h$.

Figure 6.2 shows the applied force vectors per the Mononobe-Okabe methodology.

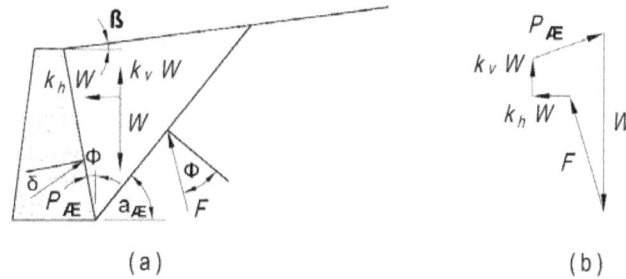

(a)   (b)

Figure 6.2 Applied force vectors on stem per Mononobe-Okabe methodology

The seismic lateral pressure diagram ($K_{AE} - K_A$) has an inverted and trapezoidal pressure diagram (See Figure 6.4) because P seismic is assumed to act at a height of 0.6 H. Stem design is based on H measured from top of footing to the retained height; for overturning and sliding H is the height through the back face of the footing along a virtual vertical plane extending from the bottom of the footing to its intersection with the backfill grade.

The K_A component has the familiar triangular distribution with P_A acting at H/3.

The height to the line of action of the combined resultant can be obtained by the equation:

$$\bar{y} = \frac{P_A\,(H/3) + (P_{AE} - P_A)\,0.6H}{P_{AE}}$$

The direction of force application, per the Coulomb equation, for a wall without a batter is assumed to be inclined at an angle (from horizontal) equal to the friction angle at the back face of the wall, δ, which is often assumed to be $\phi/2$, assuming that the wall is not battered.

Therefore, the horizontal component is:

$P_{AE\ horiz.} = \cos\left(\phi/2\right) P_{AE}$, assuming δ = $\phi/2$. See Figure 6.3.

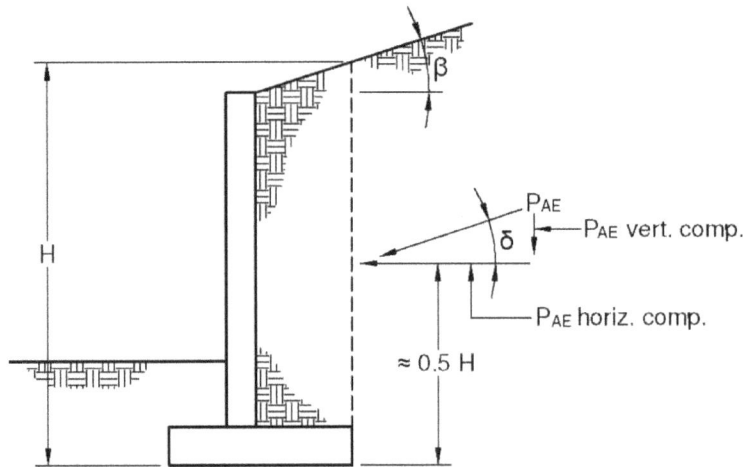

Figure 6.3 Horizontal component of seismic force

A simple approach to the design for seismic is suggested by the overlapping lateral pressure diagrams, which tend to combine into a uniform load over the height of the wall, if the height of the resultant is at or near 0.5H. Placing the resultant seismic force at a height of 0.5 H to 0.6 H, where H is the overall height of retained soil is based upon varying results from simulated tests on model retaining walls.

Therefore, $w = \dfrac{K_{AE}\,\gamma\,H^2}{2H} = 0.5\,K_{AE}\,\gamma\,H$, where w is the equivalent uniform lateral static plus seismic force. This simplification is helpful for checking moments and shears at various heights.

The inclination angle of the applied force P_{AE} if applied against the stem is generally assumed to be δ, the friction angle of the soil at the wall surface (this is unlike the Rankine equation which assumes this angle is the same angle as the backfill slope). If P_{AE} is applied to the virtual plane at the heel δ is assumed to be ϕ. The horizontal component of this force is therefore $K_{AE}\,(\cos\,\delta)$.

Designing the Stem for Seismic Forces

Of importance to the designer is the determination of the moments and shears at incremental heights along the wall. To do this it is necessary to assume the horizontal pressure distribution along the height of the wall. As stated above, a reasonable assumption is that combining the triangular static force with the trapezoidal seismic force results in a uniform load against the wall. This is illustrated in Figure 6.4.

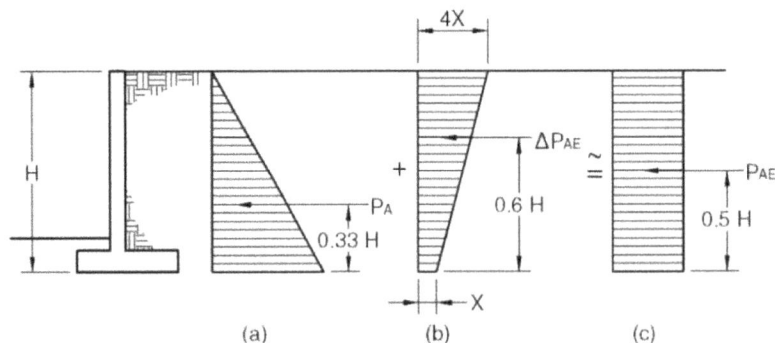

Figure 6.4 Assumption of uniform horizontal wall pressure

With this uniform load assumption the total static + dynamic force can be proportioned to the soil-retention height of the stem and the various vertical sections of the stem designed accordingly.

A recent and informative paper is *Seismic Earth Pressures on Cantilevered Retaining Structures* by Atik and Sitar in the October, 2010 *Journal of Geotechnical and Geoenvironmental Engineering,* ASCE. A Google search will yield more resources.

If it is determined that seismic design is required, there are two seismic forces that need to be considered. First, as described above, is the horizontal lateral earth pressure seismic force. Secondly, and usually not considered to act concurrently, is the seismic inertial force of the wall self-weight. The latter is primarily used for free-standing walls where there is little or no earth retention. The geotechnical consensus seems to be that these two would be in opposing synch and therefore not be applied simultaneously.

An arguable issue is whether to include the inertial force of the wall combined with seismic earth pressure. Both NAVFAC and U.S. Army Corp of Engineers appear to require this concurrent force. It seems excessively conservative. Check applicable building code requirements.

Seismic Design forces are based upon code prescribed Strength Design Load Factor, generally 1.0 factor seismic. This factored force is used for concrete components and masonry using the LRFD method. However, for overturning, sliding, soil pressure, and masonry stems, this factored force should be reduced to Allowable Strength Design (ASD) by dividing by 1.4. Note also that the factor of safety may be 1.1 when seismic is included. See IBC 2015, 1807.2.3.

Simplified Seismic Force Computation

The NEHRP has proposed a simplified approximation:

$$\Delta K_{AE} \sim (3/4) k_h \qquad \therefore \Delta P_{AE} \sim (1/2)\gamma H^2(3/4)k_h \sim (3/8)k_h\gamma H^2$$

k_h in this equation is the Peak Ground Acceleration (PGA) modified as discussed above per Provisions Sec. 7.5.1:

where $k_h = S_{DS} / 2.5$ Use $K_a - 0.34$

To verify this approximation for ΔK_{KE}, using the zip code per example above and assume level backfill slope (this simplified estimate is not applicable if there is a sloped backfill!) and phi angle of 34° and $\delta = 17°$. Insert these values, from Design Example 6, into the M-O equation and K_{AE} becomes 1.21. Then assume $k_h = 0.34$. ΔK_{AE} then becomes $(0.703 - 0.34) = 0.36$ which is close to $3/4 * 0.45 = 0.34$. This method is not applicable if there is no backfill slope and it is recommended to use the actual variables and calculating K_{AE} using the M-O equation which results in an essentially uniform applied static plus dynamic force rather than this "simplified" method.

Geotechnical Report Format

Frequently geotechnical reports will give the seismic design force as a numerical multiplier of the retained height, such as "30 H^2". This is presumably to save you time calculating K_{AE} and looking up the peak ground acceleration and following the IBC procedure to get the S_{SD} and k_h.

Here's how they obtain that value: Suppose for a particular site you calculate the K_{AE} value as 1.26 and the K_A value as 0.41. Then the seismic component of the lateral force is $(K_{AE} - K_A) \gamma H^2 / 2$. If the soil density, γ, is 110 pcf, then the soil report might say "Apply an additional seismic force of "47 H^2"[= $(1.26 - 0.41) 110 H^2 / 2$] as an inverted triangular trapezoid with resultant acting at 0.6 H".

Seismic for Stem Self-weight

When considering freestanding walls, or those with little retained height, it is necessary to determine whether wind force or seismic inertial is the more critical condition. However, an arguable issue is whether to include both the seismic force and the self-weight of the wall acting simultaneously with the lateral seismic earth pressure force. It does not appear to be defined in the codes., However, AASHTO, in 5.5.4 states: *"...seismic design forces should account for wall inertia forces in addition to the equivalent static force, where a wall supports a bridge structure..."*, but Chapter 5.6.4, referring to flexible cantilever walls, states that *"Forces resulting from wall inertia effects may be ignored in estimating the seismic lateral earth pressure"*. Judgment and code interpretation indicates whether the seismic self-weight should be applied simultaneously with the seismic lateral earth pressure.

> When the seismic force is applied when considering wall self-weight see ASCE 7-10, 12.11.1, *Design for out-of-plane forces:*
>
> $F_w = 0.4 S_{DS} I_e$ times the weight of the wall with a minimum of 10 psf.
>
> S_{DS} as defined earlier and I_e is the importance factor per ASCE 7-10, 11.5.1

Seismic Force on Non-Yielding (Restrained) Walls

Several texts (e.g. Kramer) propose the following equation (slightly revised) for restrained retaining walls with level backfill:

> $\Delta P_{eq} = \gamma k_h H^2$, acting at a resultant height of about 0.6H
>
> Where ΔP_{eq} is the added seismic lateral force, γ is the unit weight of soil, and H is the retained height.

The resultant acting at 0.6H gives a slightly trapezoidal pressure diagram, however, for ease of calculation a uniform load can be assumed, (accepting a minimum unconservative error).

It should be noted that because there are so few incidents of earthquake damage to non-yielding retaining walls – assuming that the walls were properly designed for static loads – many engineers agree that seismic design of restrained (e.g. "basement") walls may not be necessary, particularly given the factors of safety incorporated into service level design.

References

Many investigations of dynamic forces on retaining walls are reported in the technical literature. One of the most important and influential was an ASCE paper titled *Design of Earth Retaining Structures for Dynamic Loads*, by Seed and Whitman, the results from which were presented at a 1970 Cornell University conference. In that paper they also cite the pioneering studies by Mononobe and Okabe. Another contribution was a later ASCE paper by Robert Whitman titled, *Seismic Design and Behavior of Gravity Retaining Walls* in 1990. That paper considered the lateral force to be derived from an inverted triangular wedge of soil behind the wall. Seed-Whitman proposed a simplified equation, based upon the Mononobe and Okabe theory, for the combined static and seismic factor, which they termed K_{AE}, to be applied to this wedge acting against the wall. More recently a 2016 ASCE paper *Seismic Earth Pressures on Retaining Structures and Basement Walls in Cohesionless Soils* by Mikola, Candia, and Sitar. An overview article in April 2017 STRUCTURES magazine *Nonbuilding Structures* by J.G. Soules.

Basics of Stem Design

Lateral forces applied to the stem are described in Chapter 5, *Forces and loads on retaining walls*. These may include, in addition to soil pressure, wind on a projecting portion, seismic (see Chapter 6), wind, surcharges, and eccentrically applied vertical loads. Load Combinations must be considered depending upon design method (ASD or LRFD).

In the case where there is a backfill slope and an extending footing heel, it can be argued that the force applied for overturning stability, which is assumed to be applied on the vertical plane at the back of the heel extending to its intercept with the slope (see Figure 8.1), should be the force applied to the stem. However, this would significantly increase the lateral force on the stem. Another approach, and proper, is to instead use the Coulomb equation for force applied over the retained height that equation takes into account the backfill slope. Designer's choice. Consider all factors.

Tip: Here are two very rough rules-of-thumb for assuming stem thickness: If a reinforced concrete stem, try one inch of thickness for each foot of retained height, but not less than eight inches. If a masonry stem, 8" is usually adequate for walls about six feet high, and 12" for walls up to 12 feet. Higher walls, those with sloping backfills, or when surcharge load are present will require thicker stems.

The controlling design condition for reinforcement occurs at the bottom of the stem (top of footing), where the maximum stem moment occurs. Reinforcing steel must be selected to resist that moment, however, it is not economical to continue the same steel design higher up the wall where the moment is less (unless the wall is very low). Usually, after the base of the stem is designed, another design is performed several feet higher, usually at the top of the dowels projecting from the footing. At that point alternate bars can be dropped, or sizes reduced, for economy. The diagram in Figure 7.1 illustrates this concept. If the wall is very high, you may want three or four cut-off levels and perhaps a change in stem thickness, but carefully observe the influence of a battered wall on stem thickness or changes in (concrete to masonry blocks), material. See Figure 7.1.

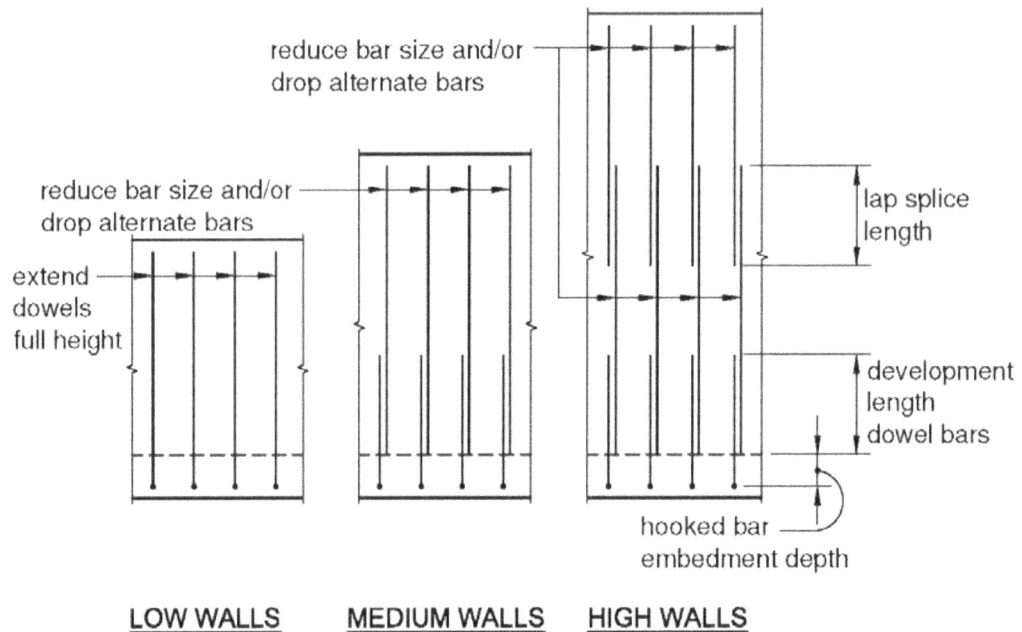

Figure 7.1. Reinforcing placement in stem

For a triangular lateral active pressure behind the wall, the moment at the base varies as the third power of the height. Thus, for example, for a 10 foot retained height the moment, provided that the proper development length is selected, at two feet above the base is 50% of that at the base. In nearly all cases the moment at the top of the dowels, provided that the proper development length is selected, is about one-half that at the base of the stem thereby halving the design requirement for continuing lapped reinforcing.

Often the stem projects above the retained height to provide a fence barrier, or a wood fence may be added to the top of the stem. In such cases, the wind load on that portion above the earth should be considered in the design, as it contributes to overturning. If the stem is essentially a yard wall and not a retaining wall and with very little earth retention, then remember that the wind can blow from either direction which will require that the wall and footing to be checked for both conditions.

Dowels from Footing into the Stem

The reinforcing at the bottom of the stem usually consists of footing bars bent up into the stem as dowel bars. Unless the wall is relatively low, say four or five feet, it is not economical to extend the dowel bars to the top of the wall, because the moment in the stem diminishes rapidly with height. Vertical dowels must only extend up to where they are not required, at which point either alternate bars can be dropped, or spliced (lapped) with lesser size bars. However, dowels must extend up into the stem a distance equal to the development length of the bar, or the required lap distance for the continuing bars, whichever is greater, provided however, that each bar extends at least 12 bar diameters beyond the point bars of that size and spacing are not needed for moment. The lap length required for the continuing bars nearly always governs. The required development length and lap lengths for both masonry and concrete are shown in the table below. Hooked bar embedments into the footing are also shown. Note the footnotes at the bottom of the Table.

Bar Size		Masonry [2] f'_m = 1500 psi		Concrete [5]		
		f_s = < 0.8 F$_S$ [2]	f_s=>0.8F$_s$<F$_S$ [3]	2000 psi	3000 psi	4000 psi
#4	L	32	48	34.9	28.5	24.7
	H[4]			13.4	11.0	9.5
#5	L	40	60	43.6	35.6	30.8
	H[4]			16.8	13.7	11.9
#6	L	48	72	52.3	42.7	37.0
	H[4]			20.1	16.4	14.2
#7	L	56	84	76.3	62.3	54.0
	H[4]			23.5	19.2	16.6
#8	L	64	96	87.2	71.2	61.7
	H[4]			26.8	21.9	19.0

(1) F$_s$ for Grade 60 = 32,000 psi per TMS 602-16.
(2) Lap length per IBC '18, 2107.2.1 = 0.002 d$_b$f$_s$
(3) For the higher f_s increase by 50% per IBC '15, 2107.2.1
(4) Assume 90° bend at bottom extending 12 inches.
(5) Per ACI 318-14, 25.4.2.2

Figure 7.2 Lap splices and hooked bar embedments

For references, this is the more complex formula for lap lengths per TMS 602, 6.1.5.1-1.

$$l_d = \frac{0.13\, d_b^2\, f_y\, \gamma}{K\sqrt{f'_m}}$$

γ = 1.0 for #3,4,5 bars, 1.3 for #6, 7, and 1.5 for #8
K = Masonry cover but not less than 5 d$_b$
d$_b$ = Bar diameter

This equation results in lap lengths greater than 48 bar diameters and has met with considerable objection. IBC 2018 modified this requirement (only for Allowable Stress Design, ASD) to l_d = 0.002 d$_b$f$_s$ but not less than 12″. This requires 48 bar diameters for Grade 60.

Horizontal Temperature / Shrinkage Reinforcing

Horizontal reinforcing is necessary to control cracks because of temperature changes and concrete shrinkage. Figure 7.3 shows minimum requirements for both concrete and masonry. There may be conditions (climate, aesthetics and better crack control) where additional reinforcement would be required at designer's option.

Typical Horizontal Rebar Spacing .0007 A_g Masonry and .002 A_g for concrete						
Mat'l	Thickness	#3	#4	#5	#6	#7
Concrete	6	9	17	18	18	—
Concrete	7	8	14	18	18	—
Concrete	8	7	12	18	18	—
Concrete	9	6	11	17	18	—
Concrete	10	5.5	10	15	18	—
Concrete	12	9	17	18	18	—
Concrete	14	8	14	18	18	—
Concrete	16	7	12	18	18	—
CMU	6	24	48	48	48	—
CMU	8	16	32	48	48	—
CMU	10	16	24	32	48	—
CMU	12	12	24	32	48	—
CMU	16	8	16	24	40	48

Figure 7.3 Horizontal temperature/shrinkage reinforcement concrete and masonry walls (inches)

The ACI 318-14 requirement for reinforcing in both faces of concrete walls over 10 inches thick is waived for retaining walls in contact with earth per interpretation of ACI 318-14, 11.7.2.3.

Resisting Shear at Stem-Footing Interface

The designer has these several options to resist shear at this interface.

One is to use a "keyway" which is a longitudinal slot "mortise" formed into the top of the footing and into which the bottom of the stem is cast. This slot can be the full width of the stem, or just the middle half. The purpose is to offer more shear resistance at the interface plane. By providing such a keyway, all or part of the shear can be resisted by compression against the side of the keyway if its depth is sufficient to resist the shear force.

However, another way of resisting shear at this interface is to consider "shear friction" across the joint. Shear friction theory considers the reinforcing steel that crosses the joint as clamping the joint together such that sliding of the joint cannot occur unless the coefficient of friction is overcome, or the reinforcing yields to allow slippage. This requires a certain amount of tension in the reinforcing must be used for this clamping force, which is in addition to tension requirements for bending design. Investigate this for an assumed condition:

$$v_u = \frac{3800}{12 \times 9.63} = 32.9 \text{ psi}$$

$$v_{allow} = 2\sqrt{f_c'} \quad \text{per ACI 318-14, 11.2.1.1}$$

$$\phi = 0.75$$

$$= .75 \times 2 \sqrt{2000} = 67.1 > 32.9 \quad \text{OK}$$

But also check shear friction available:

$$v_n = A_s f_y \mu \leftarrow \text{assume} = 0.60 \text{ coef. of friction (ACI 318-14, 11.7.2.3)}$$

$$= 0.60 \times 60,000 \times 0.60 = 21,600 > 3,800^{\#}$$

$\therefore$ OK if only consider shear friction

In this case, concrete shear is adequate, and it can be seen that tension derived shear friction offers considerable resistance if necessary.

Alternatively, you could use shear values for embedded bolts – in this case 7/8″ "bolts" which are equivalent in diameter to #7 bars at 16″ o.c. $\cong$ 3350 plf – assuming 2000 psi concrete or grout.

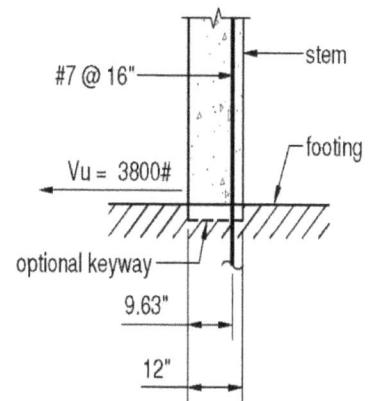

Design of Masonry Stems

The predominant building code for masonry design, and cited by IBC '18, is the *Building Code Requirements and Specifications for Masonry*. TMS 402/602.

Masonry is designed using two methods: Allowable Stress Design (ASD) and Load Resistance Factor Design (LRFD) which is Strength Design in concrete design terminology. Both are code-permitted options. Using ASD, loads are factored by 1.0, except earthquake forces are already factored, therefore to convert seismic forces to ASD divide by 1.4. Masonry allowable flexural stress is: $F_m = 0.45 f'_m$ where f_m is the compression strength of masonry.

Using load factors (LRFD) per ASCE 7-16, strength-reduction factor, ϕ, = 0.90 for flexure, 0.80 for shear, and 0.60 for bearing such as at beam supports. f'_m is typically 1500 psi and f_y for Grade 60 reinforcing steel is 60,000 psi. Refer to TMS 402/602, 9.1 for LRFD design requirements.

Concrete masonry units (CMU) are designated either lightweight, medium weight (most common) or heavy weight, and may be either solid grouted, or grouted only in those cells containing reinforcing. The vertical bars in CMU stem walls must be spaced on multiples of eight-inch centers to accommodate cell spacings. Although only cells containing reinforcement need to be grouted, it is usual to solid grout the wall. The wall weights for common combinations are shown in the Figure 7.4.

TIP: For a more specific wall weights for a wide range of conditions. See http://ncma-br.org/[pdfs/68/tek%2014-13b.pdf.

	Concrete Masonry Units											
	Lightweight 103 pcf				Medium Weight 115 pcf				Normal Weight 135 pcf			
Wall Thickness	**6″**	**8″**	**10″**	**12″**	**6″**	**8″**	**10″**	**12″**	**6″**	**8″**	**10″**	**12″**
Solid Grouted	52	75	93	118	58	78	98	124	63	84	104	133
Spacing of grouted cells — 16″ o.c.	41	60	69	88	47	63	80	94	52	66	86	103
24″ o.c.	37	55	61	79	43	58	72	85	48	61	78	94
32″ o.c.	36	52	57	74	42	55	68	80	47	58	74	89
40″ o.c.	35	50	55	71	41	53	66	77	46	56	72	86
48″ o.c.	34	49	53	69	40	45	64	75	45	55	70	83

Source: Reinforced Masonry engineering Handbook; Amrhein, 5th edition.

Figure 7.4 Weights of masonry walls, psf

Bar Position	"d" Distances for Masonry Stems				
	6″ wall	**8″ wall**	**10″ wall**	**12″ wall**	**16″ wall**
Bars in center	2.8	3.8	4.8	5.8	—
Bars at edge	—	5.25	7.25	9.0	13.0

Figure 7.5 Typical "d" distances for masonry stems

Masonry block depths are nominal block thicknesses of 6″, 8″, 10″, 12″, and rarely, 14″ or 16″. The reinforcement for six-inch walls must be placed in the center of the 6″ blocks, but for thicker walls the bars can either be centered set next to a face (termed "edge"). See Figure 7.5 for "d" dimensions for various masonry stem thicknesses. These are industry standards and assumes about 2″ minimum cover from the closest face of the wall to the centerline of the bar.

Shear and moment calculations are based upon the "effective depth, d" of the block.

When the stem thickness is reduced higher up the wall, the step should be made on the inside (earth side) so that the outside of the wall is a flush vertical surface. When stepping the wall, consideration must be given to providing sufficient lap development length for the reinforcing extending into the section below.

Minimum reinforcing of masonry stems

The code requires that the sum of the vertical and horizontal reinforcing ratios, based on the gross cross-section area A_g, be at least 0.002 and that the least in either direction be 0.0007. Spacing should not exceed 48 inches. As the principle reinforcing is always vertical, it should be at least 0.0013 times the gross cross-sectional area, and at least 0.0007 horizontally. In the latter case, #5 bars at 48″ on center or #4 bars at 32″ would suffice for an 8″ wall. Accordingly, vertical reinforcing for an 8″ wall would be a minimum of #5 at 32″ or #4 at 16″. See Figure 7.3.

Maximum reinforcing of masonry stems
It is generally not practical to exceed #8 bars at 8″ on center. TMS 402-16 limits bar size to #8.

Lap splices

For ASD design the lap splice is 0.002 d_b f_s (IBC '18, 2107.2,3). If LRFD is the selected design method, IBC '15, 2107.2.1 modifies the onerous lap splice requirement of TMS 602, equation (2-12) and replaces it with $l_d = 0.002$ d_b f_s (48 bar diameters for Grade 60 reinforcing) If f_s exceeds 80% of F_s the lap must be 50% of reactor. See Figure 6.2.

Cover requirements for reinforcing in masonry

When a face is exposed to earth or weather, 2″ cover is required for bars larger than #5 and 1-1/2″ for #5 bars and smaller. If not exposed to weather or earth the minimum cover for all bars is 1-1/2″. The industry standard "d" distances (See Figure 6.5) will provide 2″ cover.

Dowel bars into masonry stems

Footing bars bent up into a masonry stem (dowels) must extend at least the development length of the bar. Per TMS 602 this is 0.0015 d_b F_s, which, for $F_s = 32,000$ psi equals 48 bar diameters. Although arguable, this length cannot be reduced by the ratio of actual stress in the bar to its allowable stress. See Figure 7.2 for hook extension into footing. Usually use 90° bend with 12″ extention.

Stress increases for Allowable Stress Design (ASD)

Using the Alternate Basic Load Combination per IBC '18,1605.3.2 a one-third stress increase is permitted when either wind or seismic are applied. This combination also allows a reduction for seismic by 0.7 to convert to ASD.

Concrete Stem Design

Load Resistance Force Design (LRFD), also called Strength Design is generally used for concrete stem and footing design, provided all applied loads are factored per ACI 318-14 requirements: 1.2 for dead load, and 1.6 for earth pressure, wind, and live load. Use 1.4 for fluid pressure (or any well-defined density). Earthquake forces are already factored, therefore the seismic load factor is 1.0. Always check with the most recent and applicable code!

Concrete stems should be at least eight inches thick to allow space to place the reinforcing within the forms.

For LRFD design use the controlling Load Combination Factor per IBC 2018, 1605.2. See page 11.

The maximum spacing of reinforcing in a concrete wall, both vertical and horizontal is 18″ per ACI, but not more than three times the wall thickness.

Horizontal temperature reinforcing is required to be at least .002 x the gross cross sectional area of the wall (i.e. the total horizontal reinforcing should be at least equal to .002 times the wall width times its total height). If using #5 bars and larger, the ratio is .0025. If the wall is over ten inches thick temperature reinforcing is required in each face unless in contact with soil (basement walls). See ACI 318-14, 11.7.2.3. Remember that more horizontal reinforcing decreases visibility of cracks if aesthetics is an issue and may suggest reinforcing both faces.

A summary of concrete design equations is presented in Appendix A.

Concrete ultimate compressive strength is usually specified as $f_c' = 2,500$ psi. Nearly all reinforcing is now specified as ASTM, Grade 60 ($f_y = 60,000$ psi).

Minimum reinforcing of concrete stems

The minimum steel ratio $p = A_s/bd$ required per ACI 318-14, 9.6.1.2 to ensure a ductile failure of the steel is:

$$\text{min ratio} = \frac{200}{f_y} \quad (= .0033 \text{ for } f_y = 60,000 \text{ psi})$$

For example for an 8″ wall with $d = 5.5″$, #5 bars at 17″ o.c. would be the minimum required.

This minimum ratio may be waived if A_s is at least one-third greater than that required by analysis. ACI 318-14, 9.6.1.3.

Maximum reinforcing of concrete stems

The maximum amount of reinforcing steel permitted to ensure a ductile failure (tensile failure in the steel rather than brittle fracture of the concrete) is 0.75 x ρb (rho) for balanced design:

$$\rho_{bal} = 0.75\rho \quad (= 0.019 \text{ for } f_c' = 3,000 \text{ psi and } f_y = 60,000 \text{ psi})$$
Where ρ is the ratio As/bd

For an 8″ wall, $d = 5.5$, this would be #5 bars at 3″ o.c.

Determining Areas of reinforcing required

A handy equation for determining the required area of reinforcing steel, using the Strength Design method, for a given M_u is given below (taken from the *Concrete Reinforcing Steel Institute Handbook* (CRSI))

$$A_s = \frac{1.7 f_c' bd}{2f_y} - \frac{1}{2}\sqrt{\frac{2.89(f_c' bd)^2}{(f_y)^2} - \frac{6.8 f_c' bM_u}{\phi(f_y)^2}}$$

(b and d are in inches, f_c' and f_y in ksi, and M_u is factored moment in inch-kips)

using $f_c' = 3,000$ psi and $f_y = 60,000$ psi, this equation becomes:

$$A_s = 0.51d - \sqrt{.26d^2 - .0189 M_u}$$

Reinforcing cover

The cover distance for formed-concrete must be at least 2″ when exposed to earth or weather for #6 bars and larger, and 1½″ for #5 and smaller. When concrete is placed against earth, such as at the bottom of the footing, or if the wall is placed directly against earth without forming, the minimum cover is 3″.

Development length of reinforcement

The development lengths equations are in Appendix D. Note that development length can be reduced by the stress level (ratio of stress allowable divided by actual stress) in the reinforcing. Development length is an important design consideration for footing heel and toe bar development extensions and is also used to determine lapped bar splice lengths. See Figure 7.2.

Laps and splices

Where bars are spliced (lapped) the splices are classified as either Class A or B. Class of splices are defined in ACI 318-14, 25.5.2.1. To qualify as a Class A splice, if less than one-half the number of bars are spliced, then A_s for the remaining bars must be twice A_s required at the base of the stem or at the cut off height. If more than half the bars are spliced, and A_s required is more than one-half A_s provided, it is a Class B splice, requiring 1.3 λ_d. The usual case for retaining wall stems is Class B splices. Note that reduction in lap length for stress level is <u>not</u> permitted per ACI 318-14, 25.5.3.1. See Figure 6.2 and Appendix D for development and lap lengths.

Extension of dowels above footing

Projection of the dowels from the footing into the stem need only extend upward for the development strength of the bars, plus 12 bar diameters. The dowels may then lap with continuing bars of lesser size and/or increased spacing (because of a reduced moment higher up the wall). In those cases it is the lap length, not the development length that must be met. Lap lengths (discussed above) cannot be reduced by level of stress.

Drainage caveat

A common cause of distress to retaining walls is improper surface drainage allowing water to infiltrate backfill and thereby increase the lateral pressure. To relieve water pressure, "weep joints" should be provided at the lowest course in CMU walls, or by weep holes in concrete walls, which allows drainage to the outside grade. This can be done by omitting the head joint (side joint between blocks) at every other block, or 32″ on center in CMU walls. Specify gravel behind the wall to reduce the potential for the joints or weep holes to clog. Surface drainage can be controlled by using draining channels, paving, or other means. Inspect a site for irrigation sprinklers that may lead to water build-up behind a retaining wall.

Special inspection requirements for concrete and masonry

Inspection requirements for concrete walls are listed in IBC 2018, 1705.3.
Inspection requirements for masonry walls are listed in IBC 2018, 1705.4.

References

A summary of masonry design equations and code references are presented in Appendix A
An excellent reference for masonry design is James Amrhein's *Reinforced Masonry Engineering Handbook*, 5th Edition. Somewhat outdated code-wise, but good information. Another reference is *Masonry Designer's Guide, 2013,* by the Masonry Society. www.masonrysociety.org).

8.   SOIL BEARING AND STABILITY – CANTILEVERED WALLS

Overturning and Resisting Moments

The easiest way to check stability, sliding, and soil pressure, is to set up a table showing the various forces acting on the wall, together with the its moment arm measured from the lower front (toe) edge of the footing. An example of such a table is shown on Design Example #1 in Chapter 24. The tabular format provides an orderly summary of forces, moment arms and moments for easy checking of computations.

Proportioning Pointers

Here are a few pointers and guidelines to proportion the footing:

- The width of the footing for most conditions will be approximately 2/3 of the retained height.
- It is usually most advantageous to have more of the footing width on the heel side of the stem. This will put more soil weight on the heel to improve sliding and overturning resistance.
- If there is a property line on the heel side, try to get as much heel width as possible as to provide the additional soil weight. Otherwise, you will have a sliding problem.
- If you need a key for sliding resistance, try to keep its depth less than about one-fourth the retained height, but recommend not over about two feet.
- If there is a property line on the toe side, the heel of the footing may need to be wider because soil pressures are usually greater at the toe.

Overturning Moments

Overturning moments are horizontally applied forces multiplied by the moment arm from the bottom of the footing to the line of action of the force. The primary force causing overturning is the lateral earth pressure against the wall. Derived from a triangular pressure diagram, its point of application is one-third of the height above the bottom of the footing for stability calculations or from the top of the footing for designing stems. The height used to compute over-turning is on the virtual plane at the back of the footing (i.e., where this plane intersects the ground surface). Lateral pressure from a surcharge is a uniform load applied to the back of the wall, therefore its point of application is one-half the height and the moment arm is from that point down to the bottom of the footing. See Figure 5.5 which illustrates both conditions. The overturning moment from the lateral earth pressure acting against the virtual plane at the back of the footing.

Wind pressure on the stem projecting above the soil or on a fence sitting atop a wall can also cause overturning. Wind pressures are computed in accordance with the applicable building code, and generally range from 12 to 30 psf.

Seismic, if applicable, will also contribute to overturning. That was discussed in Chapter 6.

If there is significant depth of soil or ponded water above the toe of the footing, its lateral force is viewed by some as being deductible from the heel-side active force for computing overturning and sliding. Our recommendation is to disregard this concept because it may not remain in place during the design life of the wall. Only consider the depth of soil on toe side below the top of the footing when computing passive resistance.

Note that IBC 2018 removes the confusing requirement that appeared to require designers to extend active pressure all the way to the bottom of a retaining wall key. This would apply to both overturning and sliding forces, and would eliminate the additional force shown on Figure 8.1.

The general understanding is that this was only intended to require designers to *consider* that load and decide if it may have been appropriate. But it was not the intent to require this load in typical situations where there was no reason to think that the soil would ever be removed from the toe-side of the key.

The IBC 2018 section 1807.2.1 General said, "*Retaining walls shall be designed to ensure stability against overturning, sliding, excessive foundation pressure and water uplift. Where a keyway is extended below the wall base with the intent to engage passive pressure and enhance sliding stability, lateral soil pressures on both sides of the keyway shall be considered in the sliding analysis.*"

The new code section 1807.2.1 General says, "*Retaining walls shall be designed to ensure stability against overturning, sliding, excessive foundation pressure and water uplift.*"

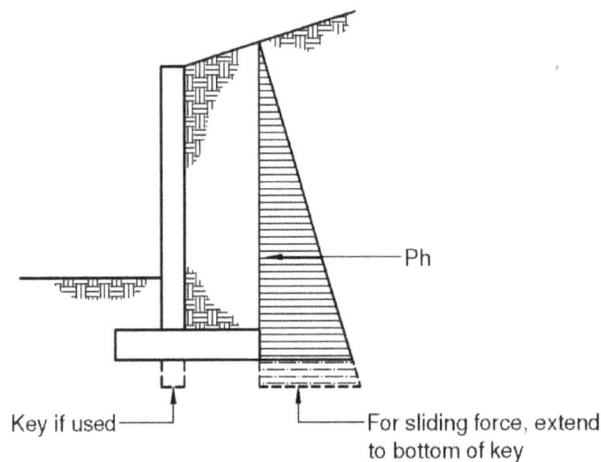

Figure 8.1 Overturning moments for a cantilevered retaining wall

Resisting Moments

By convention, resisting forces are all vertical loads applied to the footing. These forces include the stem weight, footing weight, the weight of the soil over the toe and heel, and a surcharge if applicable and any axial load applied to the top of the wall. The total resisting moment is the summation of these loads multiplied by the moment arm of each measured from the front bottom edge of the footing. See Figure 8.2.

Sliding and Overturning Safety Factors

Section 1807.2.3 in IBC 2018 clarifies that the load combinations of Section 1605 do not apply when evaluating retaining walls for sliding and overturning. Instead, it requires a load factor of 0.7 on seismic loads and factors of 1.0 on all other loads, along with an investigation with one or more of the variable loads set to zero.

The minimum factor of safety on sliding and overturning shall be 1.5, except when seismic loads are included. The minimum factor of safety on sliding and overturning shall be 1.1.

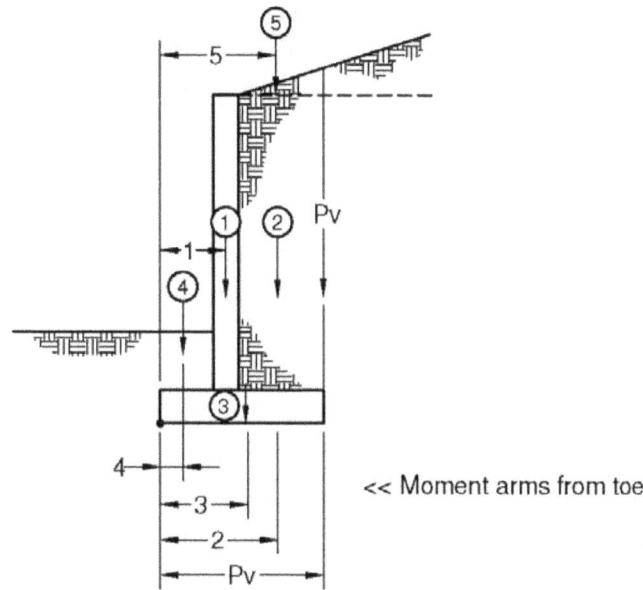

Figure 8.2 Resisting moments

To determine overturning and resisting moments, eccentricities and soil pressures, tabulate these values as illustrated on Design Example #1, Chapter 24.

Vertical Component of Active Pressure From a Sloped Backfill

If the backfill is sloped, there is a vertical component of the lateral pressure, which is assumed to act on a vertical plane at the back of the footing. This vertical component can act to resist overturning because when the wall starts to rotate there will be a frictional resistance along that plane. See Figure 8.3.

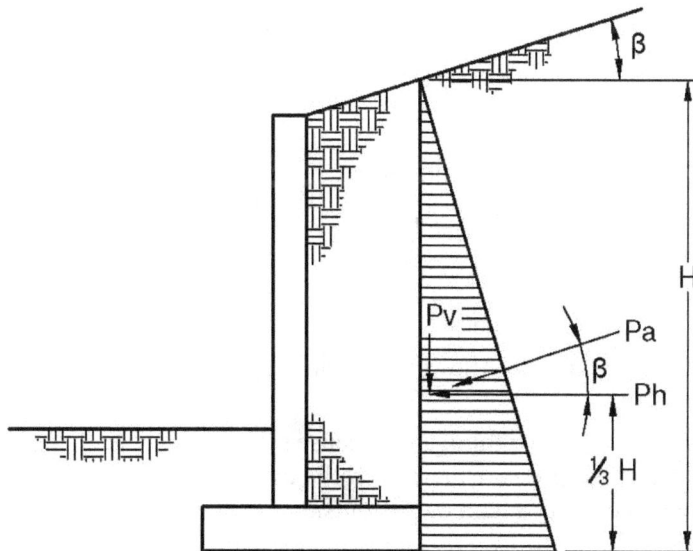

Figure 8.3 Vertical Component of Active Pressure

There is, however, controversy over whether to use this vertical component for soil pressure calculations because its use can significantly reduce soil bearing pressure and may not be justifiable if there is a large heel dimension. Similarly, it may not be justified to add vertical force to increase friction for sliding resistance. Most texts recommend using the vertical component only to resist overturning – not to reduce sliding or soil bearing. However, this judgment is left to the engineer. The concerns related to the use of P_v or that failure plane upon sliding of the wall, necessary to reach the k_A state, is an indeterminate and likely not vertical. Some value of P_V will exist but cannot with certainty be qualified. This may be one of the reasons there are so few seismic related retaining wall failures.

Determining Soil Bearing Pressure

The allowable soil bearing value, q_{all}, is within the purview of the geotechnical engineer, and usually varies from 1000 psf for poorer soil (or without a substantiating soil investigation), to 4000 psf for dense soil.

After you have assumed a footing width, taking into account property lines or other conditions that may restrict the heel or toe distances, you can determine the applied soil pressure by determining the eccentricity of the total vertical force load with respect to the centerline of the footing. This is done as follows: first determine how far from the edge of the toe the resultant vertical force acts. This is simply the total resisting moment minus the overturning moment, divided by the total vertical force.

$$x = \frac{M_{overturning} - M_{resisting}}{W}$$

W = Total vertical force (weight of concrete, soil over the heel and toe, plus loads on the soil backfill)

x = Distance from front edge of footing to resultant

Then the eccentricity is the difference between this distance and half the footing width.

$$e = \frac{ftg\ width}{2} - x$$

The eccentricity must be less than one-sixth of the footing width (that is, within the middle third) for the footing to be in contact with the soil for its full width. If this is the case, the soil pressures at toe and heel can be computed as shown in the following equation:

$$Applied\ soil\ pressure = \frac{V}{B} \pm \frac{6Ve}{B^2} \le q_{allow}$$

V = Total vertical load

B = Width of footing

e = Eccentricity

q_{allow} = Allowed soil bearing pressure

If the resultant is outside the middle third, and because soil cannot sustain "tension" between the soil and footing, the triangular pressure diagram shifts to the left as shown in Figure 8.4(c) below and becomes triangular and the resultant moves outside the middle-third. If this condition is allowed, then:

$$Applied\ soil\ Pressure. = \frac{V}{0.75B - 1.5e} = psf/foot\ of\ wall$$

May be used for the design of spread footings, where B is the width of the footing and L is its length; divide the soil pressure by L.

We recommend and some codes require that a triangular soil pressure distribution resultant "e" within the middle third of the width of the footing.

The three pressure distributions and equations are shown on Figure 8.4.

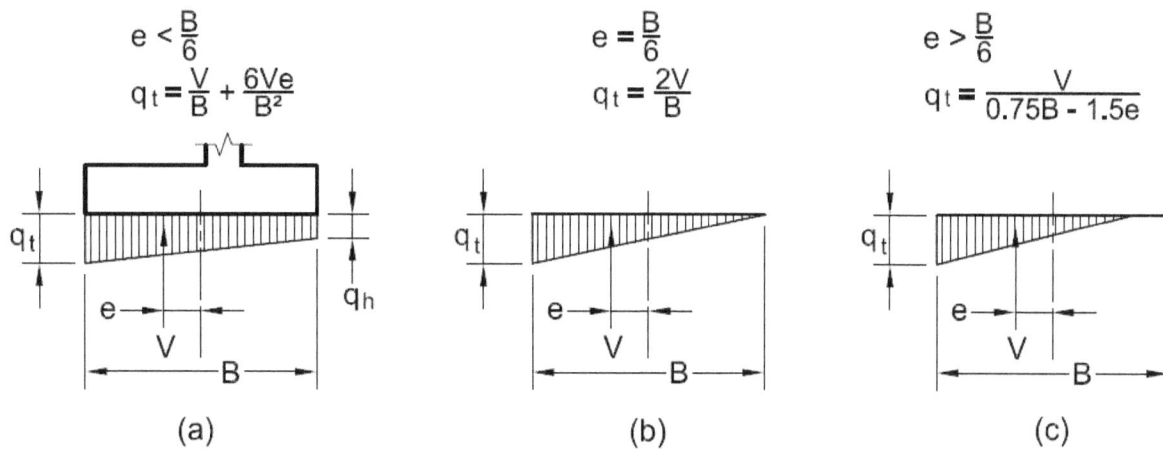

$$e < \frac{B}{6}$$
$$q_t = \frac{V}{B} + \frac{6Ve}{B^2}$$

$$e = \frac{B}{6}$$
$$q_t = \frac{2V}{B}$$

$$e > \frac{B}{6}$$
$$q_t = \frac{V}{0.75B - 1.5e}$$

(a)　　　　　(b)　　　　　(c)

Figure 8.4 Bearing pressure distribution for varying eccentricities

Ultimate Soil Bearing Capacity

The ultimate soil bearing capacity is determined by the geotechnical engineer based upon field and laboratory results and from which the soils report gives the designer allowable soil bearing values. The ultimate bearing capacity is also predominantly used for segmental retaining walls as described in Chapter 16. Ultimate soil bearing capacity can be calculated using the classical Terzahi equation:

$q_{ultimate} = CN_c + \gamma\, d\, N_q + 0.5\, \gamma\, B_e\, N_\gamma$

γ = density of underlying (in-situ) soil

d = depth of embedment of bottom of footing, ft.

B = effective bearing width, ft.

C = cohesion, psf

$N_q\, N_\gamma$ and N_c are non-dimensional coefficients per table below for usual range of soil friction angles. For their equations refer to Bowles' *Foundation Design & Analysis, Fifth Edition*, page 220. This reference also gives similar equations by Meyerhof and Hanson.

ϕ	N_q	N_c	$N\gamma$
28	17.81	31.61	15.7
30	22.46	37.16	19.7
32	28.52	44.04	27.9
34	36.50	52.64	36.0
35	41.44	57.75	42.4
36	47.16	63.53	52.0
38	61.55	77.50	80.0

Meyerhof Method

An alternate method for determining soil bearing under the footing is the Meyerhof Method, which assumes a rectangular rather than triangular stress block on the toe side. It is similar to Strength Design for concrete which assumes a rectangular stress block. For the Meyerhof Method the uniform soil pressure is the total vertical load divided by an assumed effective width of B – 2e, where B is the footing width, and e is the eccentricity of the total vertical load with respect to the footing centerline. This method results in somewhat less toe bending (and easier to compute toe moments and shears!). This method is used, and discussed, in the Segmental Retaining Wall Chapter 16. Bearing pressure distribution is shown in Figure 8.5.

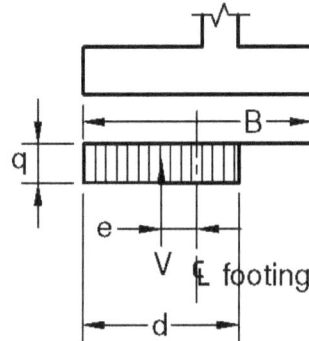

Figure 8.5 Pressure distribution for Meyerhof method

Overturning Stability

This factor with respect to overturning is the ratio of the total resisting moment to the total overturning moment, or:

$$F.S. = \frac{M_{resisting}}{M_{overturning}}$$

The minimum factor of safety with respect to overturning is 1.5 (IBC '15, 1807.2.3) with some codes and consultants advising 2.5. If seismic is included the minimum safety factor may be 1.1 per IBC cited above.

Sliding Resistance

The total horizontal driving force acting against a wall must be resisted to prevent a sliding failure. The customary minimum safety factor against sliding is 1.5, with some agencies requiring more. As for overturning, if the driving force includes seismic a safety factor of 1.1 is acceptable.

When using a key to increase sliding resistance, IBC '18 1807.2.1 reads: *"When a keyway is extended below the wall base with the intent to engage passive pressure and enhance sliding stability, lateral soil pressure on both sides of the keyway shall be considered in the sliding analysis"*. This means the wall lateral force must be calculated to the bottom of the key rather than the bottom of the footing. This can significantly increase lateral force and require design modifications. This requirement does not affect overturning which is still calculated about the base of the footing. This requirement has been rescinded in IBC ' 18.

Sliding is resisted by four ways:

Friction resistance: This is the resistance of the total vertical force (weight of wall plus surcharge) multiplied by the coefficient of friction between the base of the footing and the supporting soil. The coefficient of friction is to be determined by the geotechnical engineer, and varies from about 0.25 to 0.45. Tests have shown that actual friction coefficients are much higher, but cannot exceed tan ϕ.

Passive pressure: Passive pressure is the resistance of the soil at the toe to lateral movement from the lateral forces above the heel section. The wedge of soil in front of the wall must be pushed upward and out of the way for failure to occur. The Rankine or Coulomb equation can be used to compute the passive pressure if the angle of internal friction is known. Usually the geotechnical engineer provides this value which generally ranges from about 200 pcf to about 350 pcf. It is considered a triangular distribution, zero at the ground surface in front of the wall and maximum at the bottom of the footing or bottom of a key if applicable. However, because the soil above the footing, and in front of the toe, is usually loosely placed, its passive pressure is usually neglected, resulting in a trapezoidal passive pressure distribution starting at the top of the footing.

Both frictional resistance and passive pressure can be combined to provide lateral resistance, however, reports often limit the percentage of each which can be used in combination (e.g., 100% friction; 50% passive).

Cohesion resistance: If cohesive (silt and clay) soil or a low value of phi, then friction resistance is usually not applicable, and the cohesion (adhesion) between the bottom of the footing and soil provides lateral resistance. If this is applicable, the geotechnical report will give a value, usually around 100 psf of contact surface.

Footing keys: If the frictional and/or cohesive resistance to sliding plus the toe passive pressure resistance is not sufficient to give an acceptable 1.5 safety factor against sliding, a key can be used. A key is a trench-like deepening of the footing as shown in Figure 8.6. This additional depth to the footing is then available to further resist sliding by increasing the passive resistance. With a key, the triangular or trapezoidal passive pressure diagram extends to the bottom of the key thereby significantly increasing the passive resistance. Keys usually vary from 12 to 18 inches wide and from 12 to 24 inches or more in depth.

Design of bending stresses in the key due to passive pressure must be investigated. If the ratio of depth of a key to its width is less than about two, reinforcing is usually not required; the flexural strength of the cross section is sufficient. Compute the flexural stresses in the key. As shown below.

Assume $K_p = 2.0$

Gamma = 120 pcf

Allow. passive = 2.0 x 120 = 240 pcf

$M_{key} = (1200 - 480) \times 2' \times 1'$

$+ \dfrac{480\,x\,2}{2} \times \left(\dfrac{2}{3} \times 2\right) = 2080^{\#}$

$S = \dfrac{12\,x\,(14-2)^2}{6} = 288 \text{ in}^3$

$M_u = 1.6 \times 2080 \times 12 = 39,936^{"\#}$

$$f_r = \frac{39,936}{288} = 139 \approx 5\phi \sqrt{f_c'} \quad [ACI - 22.5.1]$$

$$= 5 \times .55 = \sqrt{2000} = 122 \text{ psi}$$

In this example, if the passive resistance of soil above the top of the footing is neglected as being loose and unreliable, then passive resistance =

$$\frac{240 \times 5^2}{2} - \frac{240 \times 1.5^2}{2} = 2730^\#/\text{ft.}$$

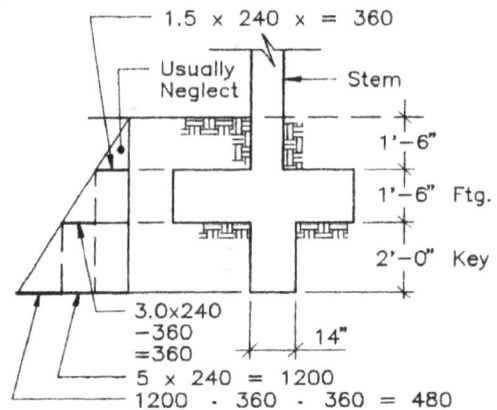

1.5 x 240 x = 360
Usually Neglect — Stem
1'-6"
1'-6" Ftg.
2'-0" Key
3.0x240 -360 =360
14"
5 x 240 = 1200
1200 . 360 . 360 = 480

Figure 8.6 Checking Concrete Flexural Stress in Key

Note also that when computing section modulus of plain concrete cast against earth, 2″ must be deducted from the depth per ACI '14, 14.5.1.7.

Deflection (Tilt) of Walls

A cantilevered wall must rotate slightly at the top to mobilize the active force in the soil wedge, implicit in the design (some texts say 0.005 times wall height). The horizontal movement at the top of the wall is the sum of the lateral deflection of the stem from the lateral forces, and the rotation of the base of the footing because of soil compression at the toe. By knowing the toe soil pressure and the k value (modulus of subgrade reaction), the settlement of the toe can be roughly estimated, and by geometry, the horizontal movement at the top of the wall determined. The soil modulus is usually in the range of 150 to 300 pci. This value must be provided by the geotechnical engineer. For typical range of values for various soil types see Figure 8.7. Tilt (lateral deflection) at the top of the wall can be approximated by the equation $\Delta_{top} = \Delta_{soil} H/B$ where Δ_{soil} is the compression of the soil based upon the bearing stress at the toe and soil modulus, H is the overall height, and B is the width of the footing.

The deflection of the stem can also be computed by conventional means using effective moment of inertia of the stem section, but this is usually not a significant component of tilt. If appearance is a concern, the front face of the wall can be battered. A rule of thumb might be a batter of $1/200^{th}$ the height of the wall (e.g. 5/8″ for a 10 foot wall).

Typical range of soil moduli are shown in Figure 8.7.

Types of Materials	Typical Values of Modulus of Subgrade Reaction, k, in lb/in^3 For Moisture Contents Shown							
	1 to 4%	5 to 8%	9 to 12%	13 To 16%	17 To 20%	21 to 24%	25 To 28%	Over 29%
Silts and Clays Liquid Limit > 50 (OH, CH, MH)	--	175	150	125	100	75	50	25
Silts and Clays Liquid Limit < 50 (OL, CL, ML)	--	200	175	150	125	100	75	50
Silty and clayey sands (SM & SC)	300	250	225	200	150	--	--	--
Gravelly sands (SW & SP)	300+	300	250	--	--	--	--	--
Silty and clayey gravels (GM & GC)	300+	300+	300	250	--	--	--	--
Gravel and sandy gravels (GW & GP)	300+	300+	--	--	--	--	--	--

Adapted from U.S. Army Corps of Engineers TM 5-809-12/AFM 88-3, Chapter 15, Table 4.1

Figure 8.7 Typical values of modulus of sub grade reaction

Global Stability

Global Stability is a term similar to "slope stability", whereby an entire soil mass under one or two tiered retaining walls slips in a rotational pattern because of low shear strength of underlying soil. The walls remain intact with this type of failure, but the soil mass slips and rotate as a bowl-shaped mass. Global stability analyses of tiered retaining walls are similar to slope stability analyses. The latter is a concern only with the weight of earth comprising of the slope. Slope stability analysis is a vast subject and numerous methods of analysis are presented in various text books.

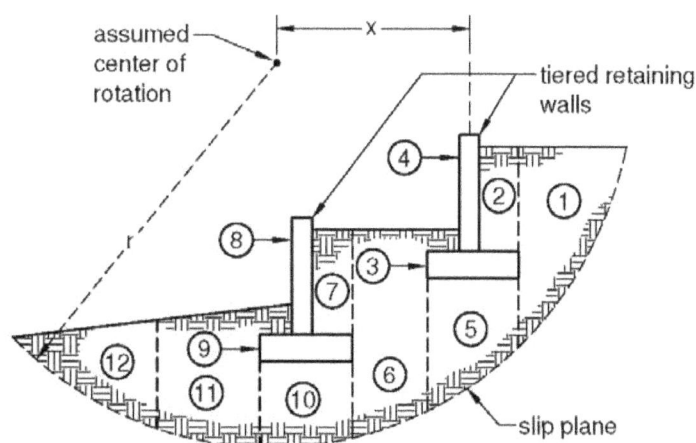

Figure 8.8 One method for global stability analysis

One method of analyzing retaining wall global stability is illustrated in Figure 8.8. This is a trial and error method to determine the most critical slip surface, which is a function of the shear strength of the soil. First, an arbitrary center of rotation and a trial slip surface are assumed and the rotational moments about it caused by the weight of the components within the soil mass and those from the weight of the walls are evaluated by summing moments about an assumed trial center of rotation, and designating the "X" distances as plus or minus. Then compare that with the moment of the unit resisting shear along the slip surface to obtain a factor safety with respect to rotation. Several trial centers of rotation and failure surfaces must be evaluated to determine the minimum factor of safety.

The above is only a generalized description of one method and the geotechnical engineer will determine whether global stability is an issue based upon an analysis of the soil underlying the proposed location of the retaining wall. If there is a potential for global failure it is the responsibility of the geotechnical engineer to recommend mitigation measures. Most geotechnical firms have software to ease the computational task.

Basics of Footing Design

The method of reinforced concrete design known as the Strength Design (SD) Method should be used to design retaining wall footings. Strength Design requires the soil pressure to be factored to compute shears and moments. See the Design Examples for procedures. Footing design based upon Strength Design requires factoring the upward soil pressure attributable to lateral earth pressure by 1.6, and pressure attributable to the weight of earth or other dead loads be factored by 1.2. Because these two components apply to footing factoring it may be reasonable to simplify by factoring the ASD soil pressure by their average: [(1.6 + 1.2) /2] = 1.4).

For footing design use factored loads to calculate the soil bearing pressure to be used for design (See Chapter 8). The soil bearing pressure is calculated by applying the factored vertical load at an eccentricity.

Use service level loads when calculating those eccentricities. Use strength level (factored) soil pressure to design footings.

Embedment of Stem Reinforcing Steel into Footing

To achieve an adequate moment connection between the stem and the footing it is customary to extend the stem reinforcing into the footing a depth sufficient to form a 90° bar hooked toward the toe (or heel if the distance is insufficient). In practice, the footing bars are placed first and extend as dowels up into the stem to lap with continuing stem reinforcing. See Figure 9.1.

This stem/dowel reinforcement must be hooked into the footing and can be bent 90° and extended to reinforce the toe. The required embedment length is specified by the following equation (see ACI 318-14, 12.5):

$$ l_{dh} = \frac{0.02\, d_b f_y}{\sqrt{f_c}} (0.7) \left(\frac{A_s \text{required}}{A_s \text{provided}} \right) $$

where d_b = bar diameter

l_{dh} = required hooked bar embedment, $8d_b$ or 6" but not less than 6 inches

Embedment depth can be reduced by the stress level in the reinforcing depending upon the application of ACI 318-14, 25.4.10.1 (d) which states that excess reinforcement can be credited except where "...*anchorage or development is not specifically required*..."

Required dimensions and radii of hooked bars are shown on Figure 9.1. Embedment requirements plus the three inches of protective concrete cover determine the minimum total depth of the footing.

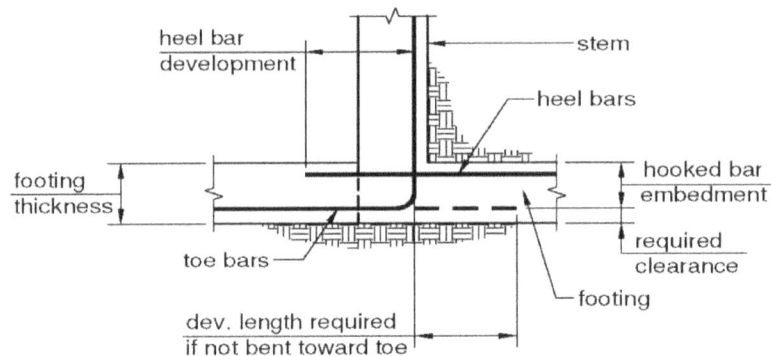

Figure 9.1 Hooked Bar Bend Requirements

If there is a key directly under the stem wall, the vertical stem reinforcing may extend down into the key for at least the development length. This will serve the dual purpose of also providing key reinforcing, if required, and will also reduce steel costs because bending of the dowels will not be required (straight bars may be used), thus reducing labor costs associated with the steel during construction.

Critical Sections

Both the toe and the heel of the footing are subjected to bending and shear forces. The critical section for bending for both the toe and heel is at the face of the concrete stem, or in the case of masonry stems the toe moment critical section is at one-quarter of the stem thickness in from the face. These moments are the sum of the upward acting moments from the soil pressure and the downward moment of the weight of soil and footing plus the influence of any surcharge loads. The critical section for maximum shear at the toe is at the "d" distance out from the face of the wall, and for the heel it is at the face of the wall.

Toe Reinforcing

As discussed above, reinforcement of the toe generally consists of the stem dowel bars bent outward toward the toe three inches above the base of the footing as shown in Figure 9.1. These bars will usually exceed the requirements for the maximum toe moment.

Depth for Shear

The allowable shear stress is $2\phi (f'_c)^{1/2}$ where $\phi = 0.75$. Divide the most critical shear required, (the larger of that at the heel or toe), by the allowable shear stress to determine the depth "d" for shear. The minimum footing thickness is then the greater of the critical shear depth or the required hooked bar embedment length plus cover of the reinforcing at the bottom. See Figure 9.2.

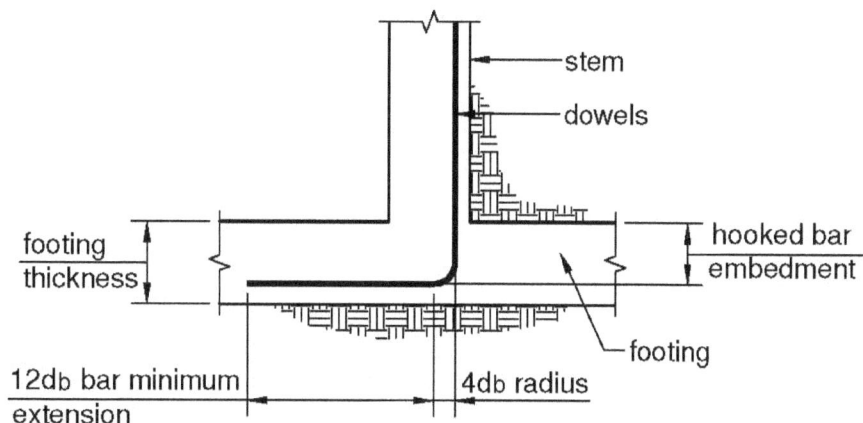

Figure 9.2 Footing Reinforcement

Heel Reinforcing Steel

Reinforcing is required to resist bending in the heel. It is calculated as for the toe steel. The critical moment occurs at the face of the stem for concrete, and inside the face of the stem for masonry stems. As discussed above, shear is computed at the face of the stem for both concrete and masonry. These bars must extend a sufficient development length past the face of the stem and are typically positioned 2 to 3 inches clear below the top of the footing.

Minimum Cover for Footing Reinforcement

The required concrete cover under bars at the bottom of the footing is three inches. At the top of the footing it is two to three inches because the top is against earth but considered a "formed" surface rather than placed against the earth.

Adequacy of Flexural Strength

If the toe or heel distance is small, less than the footing thickness, tensile reinforcing may not be required. In these cases, the flexural strength of the concrete may be adequate to resist the applied moments. When computing flexural stresses in unreinforced concrete the footing thickness used to calculate the section modulus must be reduced by two inches (ACI 318–14, 22.4.7) to allow for possible cracks.

The allowable flexural stress and shear for plain concrete (Strength Design) is:

$$f_r = 5\phi\sqrt{f'_c}$$
where $\phi = 0.60$ for flexure and shear
and $f_v = 1.33\sqrt{f'_c}$

See ACI 318-14, 9.3.5 and 22.5.1.

Although reinforcing may not be <u>theoretically</u> required, its omission is at the discretion of the engineer considering the conditions. Usually, it is wise to provide a minimum amount of reinforcing top and bottom in the footing which also facilitates the placement of temperature/shrinkage reinforcing.

Horizontal Temperature and Shrinkage Reinforcing

Stem horizontal temperature and shrinkage reinforcing was discussed earlier in the chapter on Designing the Stem. There is not a similar code requirement for footings, however, a minimum horizontal area ratio of 0.0012 is suggested. Given a 15" thick footing, this would require a #5 bar for each 18" of footing length. A minimum of two horizontal bars (longitudinal) should be provided.

Caution for High Axial Loads

If a high axial load is placed on the stem it could increase the soil bearing pressure to cause tension at the bottom of the heel rather than at the top as for the typical retaining wall. For this reversal condition the footing should be designed as a conventional spread footing with an eccentric load and thus requiring rearrangement of reinforcing.

Piles, Piers, and Caissons

Retaining walls can be supported on deep foundations; piles, pier or on caissons. These support options all perform essentially the same function: to penetrate the soil to a depth sufficient to achieve greater load bearing capacity than would be provided by a spread footing. This is achieved either by end bearing or frictional resistance along the lateral area of the shaft.

PILES accomplish this by being driven (steel, concrete, or timber) to either bear on hard strata or develop sufficient skin-friction through the depth of penetration. Concrete piles are usually the choice for retaining walls and abutments, and are either driven precast concrete, or cast-in-place in drilled bores.

CAISSONS is a term often used interchangeably with piers. Caissons are usually large diameter piers, but can have narrow shafts with a flared (bell) bottom for greater bearing area. Neither type is often used for retaining walls.

PIERS is a term used to describe a relatively short cast-in-place concrete shaft foundation. Some codes define a pier (as opposed to a pile or caisson) as having a depth-to-diameter ratio less than twelve. A pier's supporting capacity is achieved by a combination of lateral surface friction and end bearing but some codes do not allow both combined. If a masonry retaining wall has spaced pilasters, the pilasters can be cantilevered up from an embedded pier (see Pilaster Masonry Wall, Figure 18.1).

SOLDIER PILES also called Soldier Beams are piles driven at regular intervals between which timbers (lagging) are placed for temporary or permanent retainment of earth. See Chapter 21.

When to Use Piles or Piers?

The recommendation to use piles or piers to support a retaining wall will usually come from the geotechnical engineer. Conditions which would suggest using piles include poor or compressible surface or near-surface soil, the need for greater lateral resistance, space limitations when a conventional footing may be too large, or other site-specific concerns. Single-row drilled cast-in-place piers, aligned under a retaining wall, are probably more commonly used. Single rows of piers are relatively easy to install, penetrate to better soil, and resist both the vertical and lateral loads imposed from the wall above. With higher walls a double row of staggered piers is common practice. The staggering provides from greater overturning resistance and use of smaller diameter piers. Small implies diameters less than 24 inches, as opposed to large diameter piers that might be needed for overturning moment or high retaining walls.

Design Criteria

Design criteria for piers and piles is usually provided by the geotechnical engineer because IBC '15, 1803.5.5 requires a foundation investigation for deep *foundations "unless sufficient data upon which to base the design and installation is available".* This investigation generally includes: recommended type of piles or piers suitable for the site; allowable capacity curves for the various alternates, including lateral design criteria; minimum spacing; driving and installation requirements; testing requirements and related recommendation that include site-specific precautions.

To aid the geotechnical engineer, the designer should provide the total vertical load imposed by the retaining wall (weight of stem, footing, soil, surcharges, and any additional axial loads) and the total base shear (lateral force imposed by the retaining wall). Using the recommendations of the foundation investigation report the designer can then select the proper size and penetration of the pier or pile, and provide the appropriate specifications, referencing the foundation investigation report. It is important that the owner retain the geotechnical engineer to observe all aspects of the installation for conformance with the recommendations of the geotechnical report.

Pile Design

The structural design requirements for piles are covered in IBC '15 Chapters 1808 through 1812.

Lateral stability is an essential consideration for any retaining wall. To resist a lateral force piles may be either battered or the lateral force can be resisted by bending in vertically aligned piles. In the latter case, passive and active pressures can be used to determine pile/pier depth.

Consider possible site clearance problems and consult the installing subcontractor for suitability of your design when using battered piles. Generally, a batter flatter than 1(H): 4 (V) should be avoided. Combining lateral pile bending with battered pile resistance is not recommended.

Where multiple piles are used the code requires interconnected lateral restraint at the top of the piles. However for retaining walls this is achieved by the footing which also serves as the pile cap.

Pile Design Example

This example assumes the same vertical load and horizontal force as Design Example #1.

Use two rows of piles, 4 ft. apart laterally, centered under footing, and, 8 ft. on center longitudinally.

Reduce footing width to 7 ft. and increase thickness to 24", therefore footing weight the same.

V_{base} = 4,360 # $P_{vert.}$ = 9,034 #. e (eccentricity from C.L. ftg.) = 1.32 ft.

Convert to 8 ft tributary length: V_{base} = 34,880 # P_{vert} = 72,272 #

Vert. load per pile = $P = \dfrac{\Sigma V}{n} \pm \dfrac{\Sigma Md}{\Sigma d^2}$

n = number of piles (= 2).

d = distance from c.g. of piles to specific pile (= 2).

M_{ecc} = (9034 x 8)(1.32) = 95,400 ft-lbs.

$\therefore P = \dfrac{72,272}{2} \pm \dfrac{95,400 \; x \; 2}{2^2 + 2^2} = 59,986$ *ft – lbs.* max.

TIP: The above is well explained in Peck, Hansen and Thornburn Foundation Engineering, 2nd edition, 1974, pages 436, 437.

V to each pile = (4,360 x 8)/2 = 17,440 lbs.

Determine moment; pile size, and reinforcing for bending per criteria in geotechnical report.

Determine required length (penetration) of pile and lateral deflection per soil design parameters in the geotechnical report.

If impractical to resist by bending, use one battered pile on outer row, but this is generally not recommended.

V = 34,880 #

Assume batter = 1:3

∴ Axial load into pile from shear = 34,880 x $(3^2 + 1^2)^{1/2}$ = 110,300 #

Total axial load in pile = 110,300 + 91,785 = 202,285 #

Determine total penetration required for this total axial load, but include factor of safety. If loads or moments are excessive, reduce pile spacing or use an additional longitudinal row of piles or consider using large diameter piers.

TIP: Most geotechnical firms have software for pile design, such as L Pile (www.ensoftinc.com).

Pier Foundations

The consultant may recommend piers where the upper soil is weak, or where space is not available for conventional foundations. These are most commonly drilled bores, aligned in a single row under the footing which serves as a continuous pile cap, and cast-in-place concrete after the reinforcing is placed. Piers are usually spaced from six to twelve feet on center and diameters vary from a minimum required 18" to 24" or more. Spacing and diameter depend upon design requirements for sustaining both vertical and lateral loads.

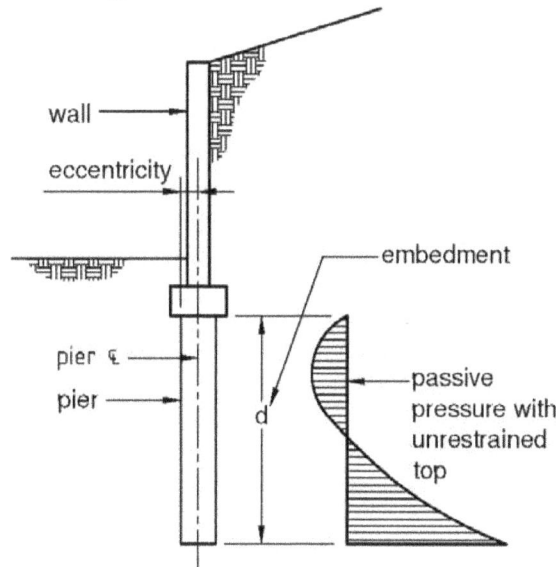

Figure 10.1 Section and force diagram for pier foundation

The geotechnical report will give recommendations and design values for end bearing values, skin friction if allowed, and permissible lateral (passive) bearing values. The geotechnical engineer may allow an increased lateral area for passive resistance, such as 1.5 times the pier diameter (usually termed the "arching factor". Creep is another factor the investigation may require, which is input as an added lateral force over a given depth of the pier. Refer to Soldier Pile Walls, Chapter 21 for additional design details for these walls.

Pile Skin Friction

There are several methods available to determine the vertical capacity of a pile or pier based upon the assumption that the load on the pile is resisted by friction between the lateral surface of the pile and the surrounding soil. The Beta method of calculating the skin friction is one of those approaches. It is applicable to driven piles in cohesive and non-cohesive soil and to cast-in-place piers in non-cohesive soil. Pile capacities for bored piers are generally set at the lower of the Beta values per table below. The friction between the pile or pier and the soil is known as the "skin friction" and is usually designated by the symbol: f_s where $f_{s,}$ = Beta * (Effective overburden stress at depth D). Beta is a coefficient that is related to soil type as indicated below:

Approximate Range of Beta-coefficients.*

Soil Type	Phi	Beta
Clay	25 – 30	0.15 – 0.35
Silt	28 – 34	0.25 – 0.50
Sand	32 – 40	0.30 – 0.90
Gravel	35 – 45	0.35 – 0.80

*Basis of Foundation Design, Bengt H. Fellenius

The pile vertical capacity is the product of the skin friction multiplied by the lateral area of the pile. In most applications the skin friction will vary with depth, usually is modeled as a step function in a mixed soil profile, but if the soil is uniform with respect to type and properties, the distribution will be triangular, varying directly with depth. In a mixed soil profile Beta-coefficients can be assigned to each soil type within the soil profile and to variations in soil properties within each segment, especially when variable density non-cohesive soil is present. The designer should review the soil profile carefully prior to assigning Beta values to various depths along the shaft of the pile or pier. The computed capacity of the pile or pier is known as the geotechnical capacity of the pile to which a suitable factor of safety is applied, usually in the range of three or more. The geotechnical capacity of a pile or pier is not the same as its structural capacity; the latter to be determined by the structural designer.

If cast-in-place piers are proposed in cohesive soil, the resistance of the pile to load is resisted by the adhesion of the clay to the lateral area of the pile. Adhesion is not the same as cohesion; adhesion being a reduced value of the cohesion. The reduction factor depends upon the consistency of the clay and the composition of the pile. Proposed values of adhesion are shown below:

Values of Adhesion

Pile Type	Clay Consistency	Cohesion	Adhesion
		Psf	**Psf**
Timber	Soft	250 – 500	250 – 480
	Med. Stiff	500 – 1000	480 – 750
	Stiff	1000 – 2000	750 – 950
Concrete of Steel	Soft	250 – 500	250 – 460
	Med. Stiff	500 – 1000	460 – 700
	Stiff	1000 – 2000	700 – 720

NAVFAC DM 7.2

An important design consideration is the maximum depth used to compute passive pressures as outlined, for example, in the IBC approach as indicated below. With this method, the embedment depth will vary depending upon whether the pier is laterally restrained at the top or unrestrained. The depth required is a function of the pier diameter, allowable passive pressures, and the applied moment and lateral shear.

Where there is not lateral restraint at the ground surface (such as a slab-on-grade), the equation per (IBC '15, 1807.3.2.1) is:

$$d = 0.5A \left[1 + \left[1 + (4.36h/A) \right]^{1/2} \right]$$

$A = 2.34P/S_1b$

b = Diameter of round footing or diagonal dimension of square footing.

d = Depth of embedment required, but not over 12 feet for use in the computation of S_1.

h = Distance in feet from ground surface to applied load

P = Applied lateral force

S_1 = Allowable lateral passive pressure per IBC '15, 1807.3.2, based upon a depth of one-third the depth of embedment, in psf. This value is usually given in the geotechnical report.

Where a moment, M_g, and shear V_g are applied, such as from a retaining wall, P becomes V_g and h becomes M_g/V_g in the above equation.

The solution of this equation requires iteration to determine "d", that is, assume a value for "d", compute S_1 and solve the equation for "d" iterate until $d_{assumed} = d_{calculated}$, usually three cycles. Note that the IBC equation is an "equilibrium statement", therefore, it is a usual practice to increase "d" by 15 to 20%, or to apply a factor of safety to the lateral passive pressure. Also, note that "b" can be increased by a factor up to 2.0 to get the "effective diameter"; consult with the geotechnical engineer.

If there is lateral restraint at the ground surface, (such as a slab-on-grade), the equation per IBC '15, 1807.3.2.2 is:

$$d = \left[4.25(M_g / S_3 b) \right]^{1/2}$$

M_g = Applied moment at ground surface, P*h as above

S_3 = Same as S_1 above, except the allowable lateral passive pressure at the full depth of embedment. This solution also requires iteration for "d".

When the diameter and depth have been determined, the next design task is to design the pier for lateral bending. An alternative to a rigorous analysis for point of contra flexure (maximum moment and zero shear), can be assumed to be one-third the depth of the pier below the ground surface. However, finite-element/spring analysis, observation tests, and practice have commonly reduced this to one-sixth the depth. With this determination the maximum design moment is obtained.

Calculating the moment capacity of a round column with bars in a spiral reinforcing configuration is highly complex, because not only are the bars at varying "d" distances from the centers of the section, but also the depth of the traditional Whitney stress block changes with the depth of the circular segment. Help for this difficulty came from an ASCE Transactions paper published in 1942 by Charles Whitney. In it he devised an equivalent rectangular section, thus vastly simplifying the calculations and reportedly being in close agreement with a rigorous analysis. In the Whitney approximation method, assume an equivalent rectangular section with total depth equal to 0.80 times the diameter of the circular column. The width is assumed to be equal to the gross circular column area divided by 0.8 times the diameter. The reinforcing is assumed to be one-half on each face, with the separating distance equal to 2/3 the diameter of the circular configuration. If compression-side reinforcing is neglected, (conservative and easier computation), then the "d" distance for design is assumed to be 0.67 times the circular diameter. This is illustrated in Figure 10.2.

Design Example using Whitney Approximation Method to Determine M_n

Assume 30" diameter; 8 - #8 bars; f_y = 60,000; f_c' = 3000 psi;, clearance = 3"; ϕ = 0.90
Gross area of circular column = $\pi \, 30^2 / 4$ = 707 sq. in.
Whitney equivalent rectangular width = (0.80 x 30) = 24"

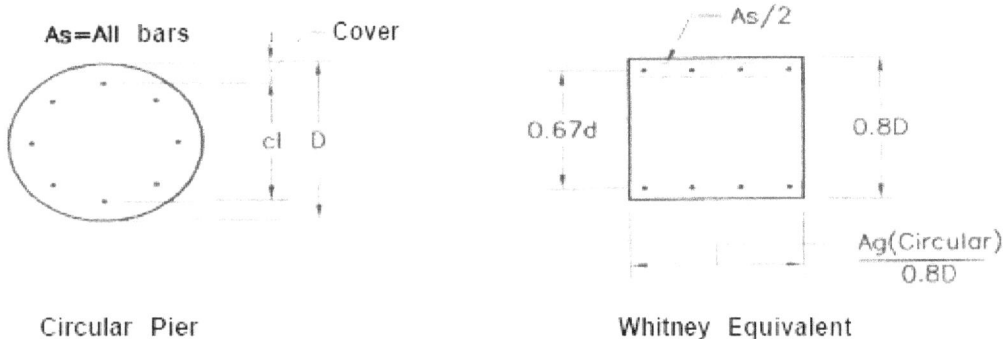

Circular Pier   Whitney Equivalent

Whitney equivalent "d" = 2/3 (30") = 20"

Figure 10.2 Whitney Approximation Method

Then use ACI 318-14 equations:

$$a = \frac{A_s f_y}{.85 f_c b} = (8 \text{ x } \tfrac{1}{2} \text{ x } 0.79 \text{ x } 60{,}000) / (0.85 \text{ x } 3000 \text{ x } 24'') = 3.1''$$

$$\phi \, M_n = 0.90 \; A_s f_y \left(d - \frac{a}{2} \right) = 0.90 \text{ x } 8 \text{ x } \tfrac{1}{2} \text{ x } 0.79 \text{ x } 60{,}000 \, [\, 20 - (3.1 \, / \, 2)] = 3248 \text{ in-kips}$$

Compare this allowable moment with applied moment, assuming applied moment is computed at the point of zero shear estimated as being 1/6 depth of pier below footing cap (assuming no lateral restraint at surface, if applicable).

Allowable shear, $\phi V_n = 0.55$ x 2 x $(f_c)^{1/2}$ x $A_g = 0.55$ x 2 x $(3000)^{1/2}$ x 20 x 24 sq in $= 28.9$ kips Axial stress can generally be ignored because it usually is less than 10% of the allowable axial stress. For example, an 8,000 psf end bearing pressure results in only 56 psi, versus an allowable of 0.25 x 3000 = 750 psi. But check if considered significant.

Between piers the footing will be subjected to torsion. The critical section for torsion is "d" from support (ACI 318-14, 9.4.4.3) which significantly reduces torsional shear required. Torsional shear is generally not a problem considering the resisting section is the footing width times its thickness. Refer to, ACI 318-14, 9.5.4.1, that allows a "threshold torsion" value below which no torsion reinforcing is required. This equation is:

$T_{th} = \sqrt{f_c^1}$ (footing in area,$)^2$ /(footing perimeter in.)

Description

The decision to use either a buttress or counterfort depends upon site restraints, such as property lines and aesthetics. A "counterfort" wall should not be confused with a "buttressed" wall. The two are different. A counterfort wall has the stiffening element on the inside of the wall, within the retained earth. See Figure 11.1. A buttress wall has the counterforts on the outside exposed side of the wall. Although most counterfort walls are cast-in-place concrete, masonry can also be used. The design procedures are essentially the same.

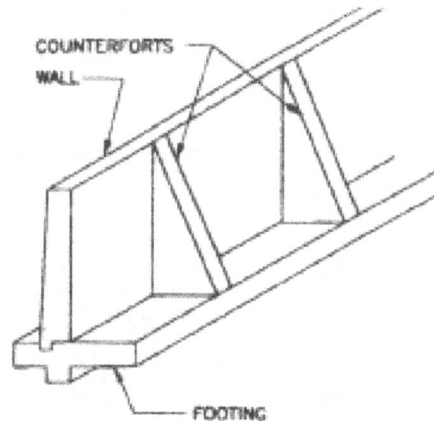

Figure 11.1 Typical counterfort wall

Proportioning

The spacing between counterforts for economical design is usually one-half to two-thirds the wall height. The width of the footing will usually be about two-thirds the wall height, or larger for surcharges or sloped backfill.

Design Overview

The design of a counterfort wall can be somewhat complex because the number of components which must be designed differently than for a conventional cantilevered wall. The steps in the design of a reinforced concrete counterfort wall are as follows (each step will be discussed later):

1. After establishing the retained height, select a spacing for the counterforts, usually one-half to two-thirds of the retained height. Determine the footing width required and the soil bearing pressures at both the toe and heel because you will need these parameters to establish the counterfort dimensions and for stability. Approach the design as if it were a continuous cantilevered wall. You can add an estimated weight of the counterforts prorated as a uniform longitudinal axial load.
2. Design the wall, as described in the following section, as a two-way slab fixed at the base and at the counterfort crossings and free at the top. See Figure 11.1.
3. Design the footing toe as a cantilever from the wall.
4. Design the heel as a longitudinal beam spanning between counterforts.
5. Design the counterfort. It will be a tapered trapezoidal shaped tension member.
6. Check the final design for stability, overturning, sliding, and soil pressures.

Designing the Wall

The wall is a two-way slab fixed at the bottom to the footing and supported (fixed) at each end where it crosses the counterforts. An assumption for vertical moment must be made based on the magnitude of the negative cantilever moment at the footing. One text (Foundation Engineering Handbook, Winterkorn/Fang, 1975) suggests (modified): $M = 0.03\, K_a\, \gamma\, H^3$ which is roughly equivalent to the fixed-end moment for a triangular loading with fixed bottom and laterally supported at top. An approximation could be made to design the wall at the cantilevered base as approximately 1/3 the moment of a pure full-height cantilever. It is suggested that this negative moment reinforcing (placed on the earth side, of course) be extended up to about one-quarter of the height of the wall, then drop or delete alternate bars.

Tables showing moments and reactions for two-way slabs with varying end conditions are presented in a publication by the Water Resources Division of the Bureau of Reclamation. Included are tables for various end conditions, span ratios, and other variables. These tables are available in pdf format from http://www.usbr.gov/pmts/hydraulics_lab/pubs/EM/EM27.pdf. Using these would give more accurate values; however, the simplified procedures described herein should be adequate for most cases.

There will be some continuity across counterforts, therefore it is suggested the horizontal reinforcing be placed in the center of the wall. Designing such horizontal reinforcing for a lateral pressure at one-half the wall height would seem prudent. Theoretically, the pressure reduces nearer the top, but it is probably practical to use the same horizontal reinforcing full height. Use your judgment to detail the reinforcing because the need for negative vertical reinforcing diminishes near the counterforts, as does horizontal reinforcing near the wall bottom.

Determining Footing Size

After the wall design is complete (its weight determined) and the spacing of counterforts known and their weight estimated (usually assume 8 inches), check a trial width of footing accounting for all forces and loads shown on Figure 11.2. Prorate the counterfort for one longitudinal length of the footing and check the footing as a unit-length conventional cantilevered wall. The heel width must be sufficient for the depth of the counterfort.

Designing the Footing Heel

The heel can be designed as a longitudinal beam spanning between counterforts, with the appropriate uniform load being the net difference between the downward weight of the soil and concrete in the heel, and the upward soil pressure. This beam can be designed as a continuous beam ($w\, L^2 / 12$) with top reinforcing between counterforts and bottom reinforcing under counterforts. If the moment is not large it may be prudent to place all reinforcing at mid-depth of the heel.

Designing the Footing Toe

The toe is designed as a cantilever from the wall, similar to a conventional non-counterfort wall, and the dowels in the stem bend outward toward the edge of the toe.

Designing the Counterfort

The counterforts are generally tapered, flaring from the top – or slightly below the top of the wall for aesthetics – to near or at the edge of the footing heel. The heel dimension will be determined by stability calculations of the counterfort (overturning, soil pressure, and sliding). Counterforts are usually 8 to 12 inches thick. The counterfort can be considered to be a vertical tapered beam with tension on the inner edge. Its applied lateral load from the retained soil will be a triangular distribution based upon the tributary area between the counterforts. The base moment and shear can be determined, and because the counterfort tapers, the moment and shear lessen higher up the counterfort, hence less reinforcing will be needed. Dowels from the footing should extend into the counterfort about three feet, therefore at that height the moment should be re-calculated and a lesser amount of reinforcing provided that would continue to the top. Anchor the tension (outer) edge of the counterfort by hooked bar embedment into the footing.

When the moment (M_u) and "d" (effective depth) distance have been determined, the following CRSI equation can be used to determine reinforcing required:

$$A_s = \frac{1.7f_c bd}{2f_y} - \frac{1}{2}\sqrt{\frac{2..89(f_c bd)^2}{\left(f_y\right)^2} - \frac{6.8f_c bM_u}{\varphi\left(f_y\right)^2}}$$

(b and d in inches, f_c and f_y is ksi, and M_u in inch-kips)

For $f_c = 3,000$ psi, and f_y - 60,000 psi, this equation becomes:

$$A_s = 0.51d - \sqrt{.26d^2 - .0189\,M_u}$$

Stability

Overturning, sliding and soil pressure calculations assume the wall and counterforts act as an integral unit, as if it is a conventional continuous cantilever wall. Include the weight of the counterforts. The overturning and resisting moments are then computed to determine safety factors and soil bearing pressures.

W_1 = Stem Weight
W_2 = Footing
W_3 = Soil over heel
W_4 = Soil in sloped backfill
W_5 = Counterfort weight / spacing
W_6 = Soil over toe
Add surcharge if applicable
Pa = Active lateral pressure
(Using Rankine or EFP)
Ph = Horizontal component of Pa
F_P = Passive pressure resistance
F_F = Friction resistance
(or cohesion)

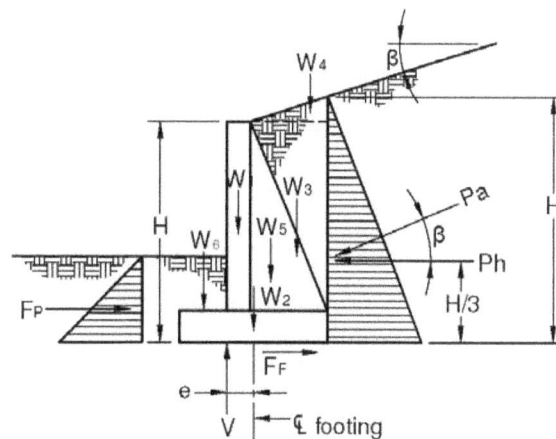

Figure 11.2 Components for force and load analysis

Description

Tilt-up concrete construction is a growing segment of the concrete industry and now accounts for over 50% of all low-rise commercial buildings and about 90% of industrial and warehouse buildings. Tilt-up yard walls, trash area enclosures, dock walls, and retaining walls are now commonplace and the use of this technique can be advantageous for retaining walls in general. This method is particularly advantageous for long walls allowing repetitive use of panels.

The primary advantage of the use of tilt-up concrete is speed of construction and the elimination of expensive formwork necessary for cast-in-place retaining walls. However, because a crane is necessary during erection, and because a casting bed is required, provision must be made for stacking panels on the site. Connections must also be made for joints between panels.

Construction Sequence

After preparing a 3″ to 4″ thick concrete casting slab (later discarded), edge forms are set, a bond breaker is sprayed onto the bed to prevent bonding of the wet concrete to the casting bed, reinforcing is placed, and the concrete for the wall is placed. To save casting area, panels can be stacked on top of each other, separated by a bond breaker, up to five or six slabs high as desired.

Unique to using tilt-up panels for free-standing or retaining walls, a trench for the foundation is first excavated and the panels set on temporary concrete setting blocks and the panel is temporarily braced. Dowels project from the bottom of the panels into the footing excavation to provide a moment connection after a continuous concrete footing is placed.

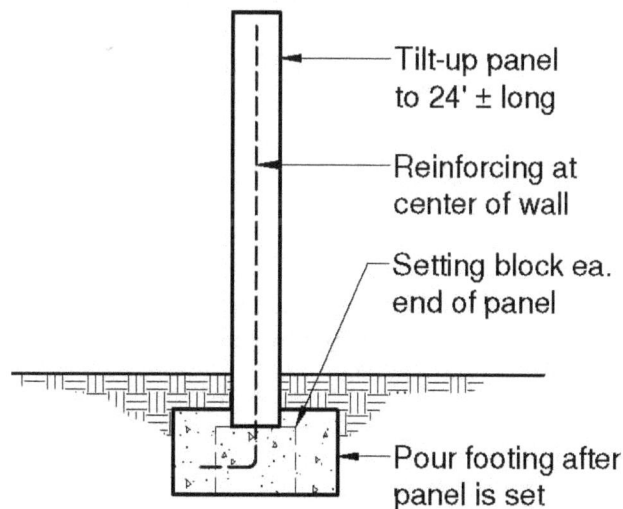

Figure 12.1 Tilt-up freestanding panel

Design Procedure

Design of the wall and foundation are the same as a cast-in-place wall. Just remember when detailing the individual panels to show the dowels projecting the proper distance out of the bottom of the each panel. The temporary setting blocks at each end of a panel remain in place and become integral with the footing as shown in Figure 12.1. Check the soil bearing pressure under the setting pads for the panel weight – it's usually reasonable to allow double the allowable bearing pressure for short-term bearing.

A low-slump concrete mix should be specified for the footing to reduce vertical shrinkage which could leave a gap under the wall. Depending on the application, it might be prudent to leave a one inch gap under the panel for dry-packing a few days after the foundation has been placed.

Free-Standing Walls

Tilt-up can also be advantageous for free-standing walls provided the length of walls justifies the use of a crane for erection. The vertical reinforcing is best placed in the center of the panel of free-standing walls because these walls are subjected to wind and seismic forces which, of course, can occur from either direction.

Foundation Design

Design the foundation width, depth, and reinforcing as for a conventional cantilevered wall.

Erecting the Panels

This type of wall is relatively low (as opposed to tilt-up panels for a building) so that the panels can be "end picked", meaning inserts are cast into the top edge of the walls, near each end, to which the lifting cables are attached. The crane then lifts (tilts) the panel free from the casting surface and, with the panel hanging plumb, caries it to its final position and lowers it onto the setting pads. Design for lifting stresses and inserts is usually done by a lifting hardware provider. They will check for tensile stress in the panel when it first lifts free (when it is a simple-span beam with bottom resting on casting slab and top supported by lifting hardware). If these concerns are understood, the design for lifting can also be done by the design engineer.

Joints between Panels

The joint between abutting panels (½″ to ¾″) is caulked for weather protection. Structural connections for longitudinal continuity are usually embedded steel plates spliced together with a field-welded overlain plate. Several per joint.

Overview

These retaining walls are usually constructed by the homeowner or a landscape contractor and rarely exceed three or four feet soil retention. These consist of wood posts embedded into the soil a sufficient depth to restrain the lateral soil pressure imposed by wood lagging spanning between the posts. The design is often based upon do-it-yourself books for wood retaining walls. Wood retaining walls are advantageous for economical construction of low walls (about five feet maximum earth retention). Such walls do require excavation into the uphill side and the low side of the wall can be used for planting. An illustration of a wood retaining wall is shown in Figure 13.1.

Figure 13.1 Typical wood retaining wall

Calculating Lateral Pressures

To design the horizontal lagging, and the cantilevered support posts, the lateral soil pressure can be determined using the Rankine equation as described in Chapter 6. For example, the lateral pressure at a depth of three feet, with a soil density of 110 pcf, phi angle of 34°, and a level backfill, would be 94 psf acting horizontally at the lower most lagging. The lagging at that depth would be designed for that lateral force along the entire span between posts. The code prescribed minimum lateral earth pressure for a level backfill is 30 psf/ft, but increases if a sloped backfill and for different soil types. See IBC '15, Table 1610.1.

Lagging Design

Lagging usually consists of planks with a nominal thickness of 2″, 3″, 4″, or 6″. Thicker planks are generally not economical but may be necessary for higher walls. Allowable stresses are based upon the species selection and given in National Design Standards for Wood Construction (NDS), 2015 Edition. All stresses should be based upon long-term loading and wet conditions of use. Spacing between posts is usually determined by how far a 3″ plank for a given depth will span. For the above example, a 3″ x 12″ plank (dressed dimensions 2-1/2″ x 11-1/4″) has a section modulus (weak axis) of 11.7. in^3. Assuming an allowable stress of 900 psi this plank could safely span about 8-1/2 feet based upon 94 psf lateral pressure. The permitted spacing of the lowest plank would determine the maximum post spacing. When placing lagging a space should be left between planks of one-half inch to facilitate drainage – water saturated soil behind the wall should be avoided because it significantly increase lateral pressure. Lagging is usually connected to the posts by galvanized lag bolts or spikes. Often the lower planks are double layered construction to allow the use of the same thickness of lumber throughout. Treated lumbar is recommended but otherwise be sure to use a preservative.

Lagging deflection should be checked for aesthetics.

Post Design

Wide selections of wood posts are available and the determination is based upon economy and availability. Wood posts can be 6″ x 6″, 6″ x 8″ (usually oriented flat for better embedment area and fastening from lagging) or 8″ x 8″. Also, railroad ties (usually 6″ x 8″) or round poles such as available at home supply outlets. Telephone poles, which are generally tapered 12″ to 16″, are also used. If posts are pressure treated to extend lifespan a reduced fiber stress may be appropriate. Use judgment.

When the tributary lateral force is determined, the selection of post or pile type and size can be made, then the required depth of embedment calculated for cantilever stability.

The posts can either be embedded in drilled holes and backfilled with gravel, or encased in lean concrete. The advantage of the latter is permanency and more passive pressure area to reduce embedment depth.

The required embedment depth can be determined by the building code equations (see IBC 2018, 1807.3.2.1) shown on page 74. The allowable passive pressure in psf/ft, per IBC 2018, Table 1806.2 is 150 psf per foot of embedment depth for sand, silty sand, clayey sand, and silt and sandy silt; if higher clay content the allowable is 100 psf/ft.

Example construction section and force diagram is shown in Figure 13.2 and connection details are shown on Figure 13.3.

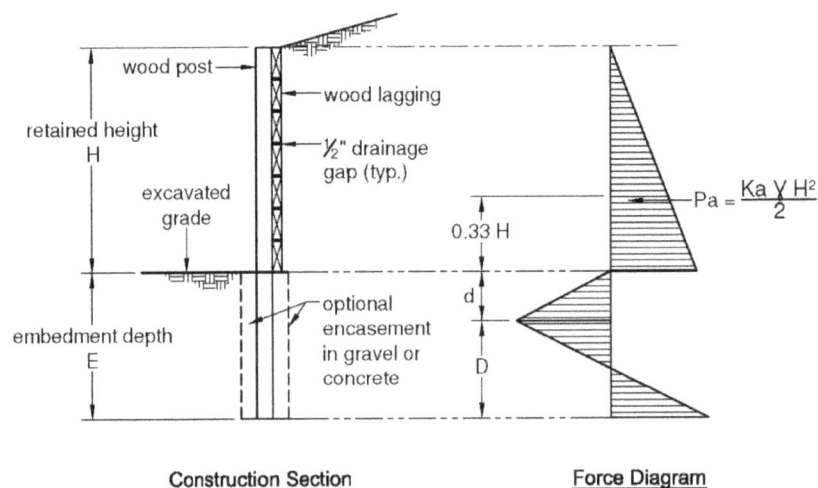

$$Pa = \frac{Ka \, \gamma \, H^2}{2}$$

Construction Section   Force Diagram

Figure 13.2 Section of a typical wood retaining wall

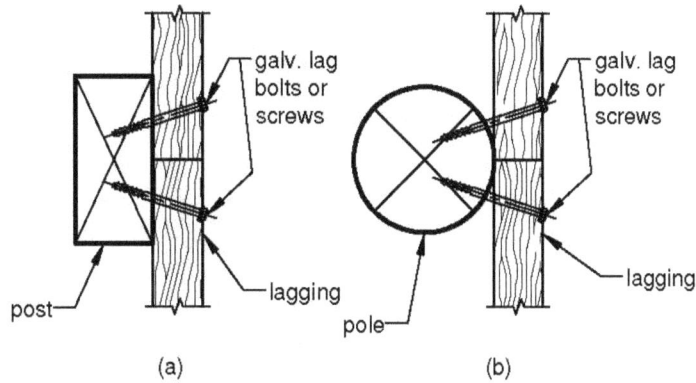

Figure 13.3 Sample connection details

Summary and Caveats

Although economy and aesthetics may lead to the choice of a wood retaining wall, it lacks the durability of a masonry or concrete wall and therefore thought should be given for the best solution for a given site. If questionable soil or unusual site conditions exist, a geotechnical engineer's advice should be requested. Inadequate depth of embedment of the posts can lead to tilted or bulged walls. Proper design of the lagging and embedment are essential as well as aesthetic considerations.

Overview

Gravity walls depend upon bulk weight for stability, as opposed to a cantilevered retaining wall fixed to a foundation. Some of the many types of gravity retaining walls were described in Chapter 1. Most gravity retaining walls are relatively low, such as used in landscaping, and do not require engineering per se – the design is intuitive to the astute builder. Most landscaping walls do not have a footing, rather are founded on a gravel base.

Note that retaining walls not over four feet from bottom of footing to retained height, and if without a surcharge, do not require a building permit per IBC '15, 105.2(4).

Gabion walls, crib walls, and large-block gravity walls are discussed in Chapter 15.

The design of the more common types of gravity walls composed of rubble, stones, and mass concrete are discussed in this chapter.

Design Procedure

The design of a gravity retaining wall of concrete or bonded (mortar/grout) stone involves seven basic steps:

1. Calculate the dead weight of the wall, including all components and any superimposed surcharge or axial load, plus tributary earth weight over the base.

2. Based upon (1) compute the resisting moment about the front edge of the base.

3. Determine the lateral soil force and its line of action. The Coulomb Equation (see Chapter 5) should be used because it includes backfill slope, batter of the wall, and the soil friction angle at the wall interface. You may consider the use of the vertical component of the active pressure, which is assumed to act vertically at the back edge of the wall footing. The line of action for the resultant lateral force is assumed to be the wall friction angle plus the inclination angle of the wall batter. Alternatively use the Rankine equation with the force diagram in Figure 14.1.

4. Check stability by computing overturning moment, resisting moment (per above), and determine factor of safety (1.5 minimum).

5. Check that soil bearing is within code allowable or geological report recommendations.

6. Check sliding. Coefficient of friction is generally 0.25 to 0.40. If soil is clay, cohesion would control.

7. Verify that little or no flexural tension exists in the wall. Check at several locations by calculating the section modulus of the wall and lateral moment at each selected height.

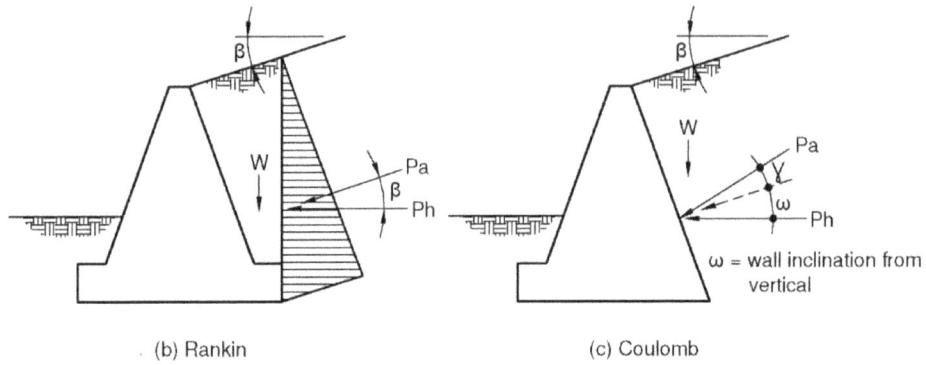

(b) Rankin   (c) Coulomb

Figure 14.1 Force Diagram for Concrete Gravity Wall

For concrete gravity walls some reinforcing is advisable for crack control. ACI 318-14 requires $0.002A_{gross}$ minimum horizontal reinforcing for walls.

For example calculations for a gravity retaining wall see Design Example 9, Chapter 24.

Descriptions

Gabion walls consist of steel wire baskets filled with rock and stacked as units to form gravity retaining walls. Similar baskets have been used since ancient times and the word "gabion" does not refer to an inventor but rather to Italian and Latin words meaning "cage". Today, the cages are manufactured, generally, in three foot by three foot by three foot steel wire panel sides which at the job site are unfolded to form a cage, which are filled with rock, tied together, and assembled into the retaining walls. Since mesh openings are generally 3 inches square, the rock infill should be 3 inch to 8 inch clean hard stone. Perpendicular to the plane of the wall the wythes can be 1, 2, 3 or more units deep and can be stacked in successive courses to a height usually not more than about 15 feet.

Note: The masonry term "wythe" means one vertical section of wall one unit in thickness.

Similar in concept, precast large concrete blocks, which are commercially available from a number of vendors and concrete plants, can be laid one or more blocks deep (wythes) and stacked to retain soil to 12 feet or more. Such blocks can be laid with the front exposed side flush or with successive blocks stepped back.

If the front face is flush, it is customarily tilted into the soil about 6° for aesthetics.

Design Methodology

The cages are wired together and because of their mass are considered one rigid cohesive mass for design purposes. Gabion walls are designed or analyzed in the same manner as gravity walls. Resisting moments are taken about the front lower corner of the first row and overturning moments are applied to the back face using the Coulomb method for calculating K_a. Density of the gabion units is usually taken as 120 pcf. Refer to Figure 15.1 for conceptual example of a flush-face wall.

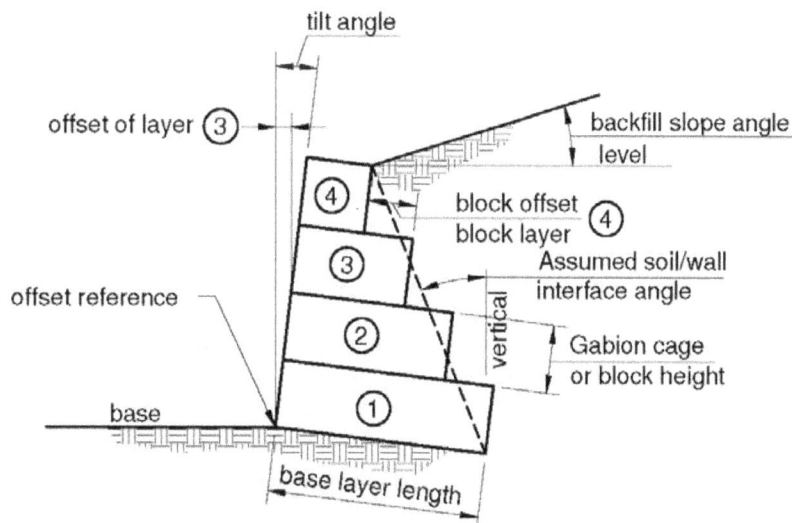

Figure 15.1 Example of gabion wall analysis section

Lateral pressures are computed by the Coulomb equation shown in Figure 15.2.
Sliding resistance is the ratio of total weight of the wall multiplied by the wall-soil friction factor,
and, divided by the total lateral thrust. This ratio should be at least 1.5.

$$K_a = \frac{\sin^2(\alpha + \phi)}{\sin^2 \alpha \sin(\alpha - \delta)\left[1 + \sqrt{\dfrac{\sin(\phi + \delta)\sin(\phi - \beta)}{\sin(\alpha - \delta)\sin(\alpha + \beta)}}\right]^2}$$

β = Angle of backfill slope
φ = Angle of internal friction
α = 90° + (tilt angle) – (soil-wall interface angle as shown in Figure 15.1).
δ = Angle of friction between soil and wall (usually assumed to be 2/3φ to 1/2/φ)

Figure 15.2 Coulomb equation

When the Coulumb equation is used to compute lateral pressure, the α angle to insert is 90° +
(positive tilt angle) – (assumed soil/wall interface angle per Figure 15.1).

The total lateral force, $Pa = K_a * \gamma * H^2 * 0.5$, where H is the vertical retained height adjusted for
inclination if applicable. Hence the horizontal component of P_a is cos [(δ ± (tilt angle of wall
from verticals)] .

Overturning moment resistance is simply the weight of each basket or block multiplied by its
moment arm from the front reference point edge to its center of gravity on a per foot of wall
basis. Successive stacked courses are added and accumulated to obtain a total overturning
resisting moment.

Foundation Pressures

Gabion walls generally do not have a concrete footing but rather are set on a firm level base,
often gravel.

To compute soil bearing value knowing the resisting moment and overturning moment the
following equation can used for determining eccentricity from the center of the mass. This
eccentricity should be within the middle third of the base width.

$$e = B/2 - [(M_r - M_o) / V]$$

e – eccentricity; B = base width; M_r = resisting moment; M_o = overturning moment;
V = vertical load

Resultant is within the middle third of the footing if e ≤ B/6, then the soil bearing pressure is:

$$\text{Soil bearing toe and heel} = (V / B) \pm (6*V*e / B^2)$$

If resultant is outside the middle third (not recommended), and since there can be no tension at
footing soil interface, the soil bearing becomes:

$$\text{Soil Bearing at toe} = V / (0.75*B - 1.5*e)$$

Sliding

Sliding on the base must also be checked. Sliding safety factor (1.5 or 2.0) = V μ / P_a(horiz), where μ is the coefficient of friction at the base-soil interface, usually ranging from 0.25 to 0.45.

Seismic Design

Seismic design – if applicable – is similar to the discussion for segmental walls in the next chapter.

Gabion Walls Using Mechanically Stabilized Earth

Although gabion and large block walls can be stacked to accommodate considerable retained heights, conditions may warrant increasing their capability by using horizontal layers of geogrids, or similar mats, embedded between block layers and extending back into the soil beyond the theoretical failure plane in the backfill to achieve an integral soil mass. Termed mechanically stabilized earth (MSE), this concept and the design procedures are discussed in the next chapter on segmental retaining walls.

References

Few textbooks discuss gabion walls. The best source of information is from vendor literature. For example: http://www.gabions.net/technical.html.

Note that although this chapter primarily deals with "gabion" walls the same design methodology may be used for precast concrete blocks stacked in nearly any configuration.

Overview

Segmental block retaining walls (SRWs) are composed of dry-stacked masonry blocks usually manufactured as proprietary products. They have gained wide acceptance for high earth retention condition and are seen everywhere: leaning against hillsides alongside highways, behind shopping centers, providing tiered grade changes for developments, highways, railroads, bridge abutments and other applications. See Figure 16.1.

Figure 16.1 Geogrids reinforced SRW wall

Advantages include relatively fast construction; as the footing consists of a gravel levelling pad, and the units are dry-stacked without mortar, steel reinforcing, or grouting. The designer has a choice of block sizes, textures, colors and configurations, from a variety of vendors. Retained heights of 40 feet or more can be achieved using tie-backs geogrids far exceeding economical limits of conventional masonry or concrete retaining walls. These do, however, have limitations. If a segmental retaining wall requires geogrids for stability, this requires an available space behind the wall of approximately 70% or greater of the wall height for the placement of the geogrid reinforcement layers. If space is unavailable, a segmental wall is not an option. Buried utility lines or drain lines in the backfill zone may also be constraints for a segmental block wall. The segmental blocks as a fascia can also be used for soil nailed walls or soldier beam and wood-lagging walls.

Segmental walls are of two types: pure gravity walls where stability depends solely upon the resisting moment of the stacked blocks to exceed the overturning moment of the lateral soil pressure. This stability problem limits the height to less than four feet, although some vendors offer larger blocks enabling greater retained heights.

Higher walls, the more common type of segmental walls use layers of geogrids placed in the backfill for soil tie-backs as the wall is constructed. This results in a mass of reinforced soil (also termed Mechanically Stabilized Earth, MSE) which can be used en masse to improve resistance to overturning and sliding. To be effective, each layer must be properly connected to the block facing by engaging the geogrid within block joints, and extending behind the wall and beyond the failure plane a distance sufficient for anchorage. The vertical separation between geogrid layers is usually two- to three blocks, but varies with design requirements. The length of the reinforced zone is usually a minimum of 60% to 70% of the wall height.

Many engineers specialize in designing segmental retaining walls. Their design can be quite complex, particularly for higher walls using geogrids, and other improvements that are to be accommodated in the geogrid reinforced zone, such as caissons, drain liners on shallow foundations for light weight structures. Consultation with a selected block vendor is recommended and many offer design software.

Segmental Blocks

Segmental Blocks are concrete blocks with compressive strength of 3,000 psi or greater, and, in the US, they are manufactured per proprietary designs at licensed local plants. The blocks come in many choices of texture, color, sizes, and configurations. The blocks vary in size, with the most commonly used blocks being 8-inch high with depths varying from 10″ to 24″. The block width for the most commonly used blocks is 18 inches. Blocks with dimensions smaller than these are available for non-engineered landscape applications for retaining heights of about three feet or less. All of these blocks weigh between 30 and 110 lbs each. So called "big blocks" are also available from some vendors, weighing two tons or more and placed by small cranes.

The blocks are designed to allow construction of walls with vertical batter -- angle of the wall face to the vertical -- to as much as over 15 degrees from vertical. To control batter most segmental blocks have offset lips or other means, such as pins between units, to control the offsets as successive courses of blocks are placed.

Angle of wall batter = $\tan^{-1}$ [(offset per block) / (block height)]

Some blocks have interior voids which are infilled with granular backfill material. Weight per square foot of wall surface is often assumed to be 130 pcf for both block weight and infill.

All vendors have web sites for more information and technical data. Best source: a Google search for "segmental retaining walls".

Segmental Gravity Wall Design

For segmental gravity walls to be stable, the resisting moment should exceed the overturning moment by a factor of safety of at least 1.5. This limits the height of gravity segmental walls to about four feet, depending upon the batter of the wall and block type. For larger blocks that are in the market, the gravity wall height can be greater.

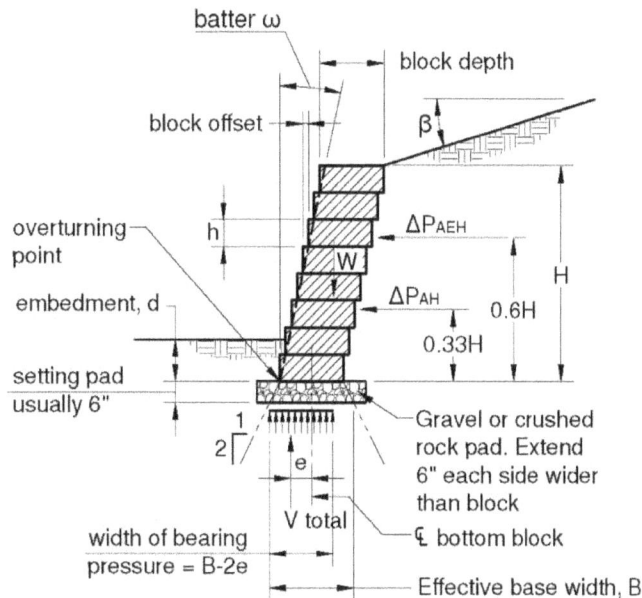

Figure 16.2 Forces on gravity SRW walls

The design procedure for gravity walls follows these steps:

Select the block vendor for texture, color, size and configuration desired. This is often dependent upon proximity to distributors.

1. Determine the retained height required and embedment depth below grade. Embedment depth is usually one block course or one foot. Total wall height is equal to the full retaining height, plus the embedment.

2. Determine surcharges, if applicable, from backfill slope or roadway traffic. If seismic design is required see below for seismic design.

3. Check "hinge height", which is the height to which blocks can be stacked, with offsets, before tipping over. The equation for this is:

    Hinge height = (block depth) / [(tan (batter angle)]

    Don't stack higher than this or the wall may collapse!

4. Determine soil properties: Soil unit weight and internal friction angle (phi) for both internal (backfill soil) and external (in-situ, or natural) soil. Backfill should be granular soils, with phi angle greater than 30 °. Ideally it would be USCS Group GW (well-graded gravels, gravel-sand mixtures, little or no fines, per Unified Soil Classification System – see Appendix B). However, for economic reasons, on-soils such as, clayey sand, silty sand, and sand are also used, provided their phi angle is considered in the design and compacted property under monitoring and testing by Project Geotechnical Engineer.

Check lateral soil pressures

Calculate coefficient of active pressure, K_{ah} (horizontal component). Use the Coulomb equation because it accounts for the friction angle at soil-wall interface and the batter angle. The friction soil-wall interface angle is usually assumed to be 2/3 phi (backfill soil). The batter angle is determined by the block-to-block offsets and is equal to $\tan^{-1}$ [(offset per block) / (block height)].

The Coulomb equation

$$K_a = \frac{\sin^2(\alpha + \phi)}{\sin^2 \alpha \sin(\alpha - \delta)\left[1 + \sqrt{\dfrac{\sin(\phi + \delta)\sin(\phi - \beta)}{\sin(\alpha - \delta)\sin(\alpha + \beta)}}\right]^2}$$

$$K_a \text{ (horiz.)} = K_a \cos\delta$$

β = Angle of backfill slope
ϕ = Angle of internal friction
α = Wall slope angle from horizontal (90° + batter angle from vertical)
δ = Angle of friction between soil and wall (usually assumed to be $2/3\phi$ to $1/2/\phi$)

Check inter-block shear

The shear at any depth "z" = K_a(horiz) [γ z^2 0.50 + (D + L) z]

where γ = backfill soil density and D and/or L = Dead load or Live load.

The maximum interface shear will be at the lowest joint. The shear resistance will be the weight of the blocks above which compresses this joint ("N" value) inserted into the vendor's tested shear resistance equation.

Check sliding

The total sliding force is the shear at the base of the wall. This is the resistance offered by the coefficient of friction between lowest block and the gravel or levelling pad, or the friction between the levelling pad and in-situ soil below. This is generally given by the equation R = N tan ϕ_e, where R is the resistance available, N is the weight above, and ϕ_e is the friction angle of the base (in-situ) soil – often assumed to be 40°. The safety factor against sliding should be at least 1.5.

Check overturning

Overturning moment at any depth "z" = K_a(horiz) [γ z^3 0.17 + (D + L) z^2 0.5].

The resisting moment = 0.5 N (t + H * tan ω), where N = weight of block stack.

H = height of wall, t = depth of blocks, ω = wall batter angle from vertical.

If more than one wythe, adjust accordingly.

The overturning ratio (resisting moment / overturning moment) should be at least 2.0 per NCMA.

Check soil bearing pressure

For SRW walls the Meyerhof Method is used to determine bearing pressure. This assumes a rectangular pressure distribution under the footing, as opposed to a triangular distribution. The total vertical force is assumed to be distributed uniformly over an effective base width. The effective base width is less than the full width by a distance equal to twice the eccentricity of the imposed load on the full footing width (easily verified with a diagram).

e = [(base width) / 2] – [[(resisting moment) – (overturning moment)] / (total vertical load)].

B_e = effective bearing width = B – 2e, where B is the total bearing width.

Soil bearing capacity

Ultimate soil baring capacity can be calculated using the classical Terzaghi equation:

γ = density of underlying (in-situ) soil

d = depth of embedment of bottom block, ft.

B_e = effective bearing width, ft. (see above for methodology)

$Q_{ultimate} = \gamma \, d \, N_q + 0.5 \, \gamma \, B_{e,} N_\gamma$ (The term N_S is omitted because cohesion is not considered in design.)

N_q and N_γ are non-dimensional coefficients is in per table below for usual range of soil friction angles. For their equations refer to Bowles' *Foundation Design & Analysis, Fifth Edition*, page 220.

ϕ_i	N_q	N_γ
30	18.4	22.4
31	20.63	26.0
32	23.2	30.2
33	26.1	35.2
34	29.4	41.1
35	33.3	48.0
36	37.8	56.3

Seismic design -- gravity walls

Seismic design for segmental gravity walls would rarely be required because of the relatively low height and exemption from most codes. Depending upon the location and local building codes, seismic design may not be required, generally for walls up to 6' height.

If seismic design is required, two components must be considered: seismic force from earth pressure and seismic force from wall inertia. The former is computed using the modified Coulomb equation below, and the latter uses the k_h factor applied to the wall mass.

K_{AEint} = active earth pressure coefficient, static + seismic (NCMA format)

$$= \frac{\cos^2 (\phi_i + \varpi - \theta_{int})}{\cos \theta_{int} \cos^2 \varpi \cos (\delta_i + \varpi + \theta_{int}) \left[1 + \sqrt{\dfrac{\sin (\phi_i + \delta_i) \sin (\phi_i - \beta_{int} - \theta_{int})}{\cos (\delta_i + \varpi + \theta_{int}) \cos (\varpi + \beta_{int})}}\right]^2}$$

Where θ seismic inertia angle = $\tan^{-1} K_h$, α = wall slope to horiz. (90° for a vertical face), ϕ = angle of internal friction, β = backfill slope, δ = wall friction angle, and ω = wall batter angle.

The horizontal component is $K_{AE} \cos \delta$.

Added seismic force = $(K_{AE} - K_a) \gamma H^2 0.5 + (K_{AE} - K_a) D H + k_h w H$

D = dead load surcharge if applicable; H = height of wall; w = weight of wall psf; and γ = soil density
Where $K_{HW}H$ is the inverted weight of the wall.

The seismic component is usually designated as $\Delta K_{AE} = (K_{AE} - K_A)$

The seismic earth component is assumed to act at 0.5 H and the dead load and wall inertial forces at 0.5 H.

The value of k_h is usually assumed to not exceed 0.15, however, NCMA states k_h should be one-half the peak ground acceleration (PGA). In high seismic areas this results in a seismic force for which most gravity walls could not resist. For example, in Newport Beach, CA, the PGA is about 0.58, which equates to a k_h of 0.29. The acceleration can also be equal to S_{DS} / 2.5, refer to Chapter 6.

If seismic forces are included, the safety factor for sliding and overturning can be reduced to 1.1.

Reinforced Earth - Geogrid Wall Design

The soil retaining height of a segmental wall can be increased by placing successive layers of geogrids in the backfill as it is being placed from the face of the blocks. This results in a composite mass of "reinforced earth" behind the wall (also called Mechanically Stabilized Earth, MSE) which acts in en masse to resist overturning and sliding. This enables segmental walls to reach retaining heights of forty feet or more.

Forces acting on a segmental wall with geogrids is shown in Figure 16.3. The 3rd edition of NCMA's design manual recommends a rectangular seismic force distribution.

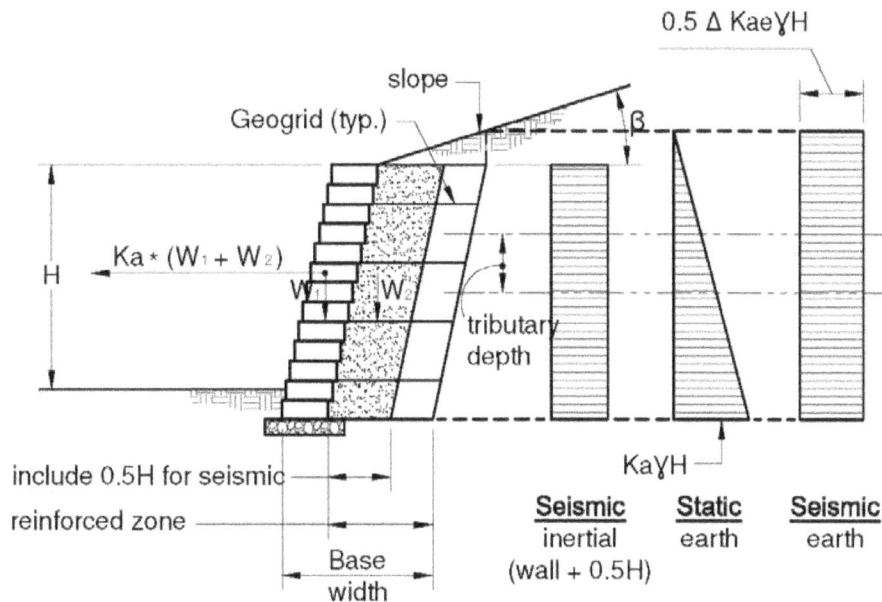

Figure 16.3 SRW with Geogrids

Construction sequence

Construction begins with an excavation behind the wall extending a distance determined by design, but usually a minimum of 60% of the height of the wall. A gravel or crushed stone leveling pad is used as a base for the segmental units. This leveling base is usually six inches thick and extends a minimum of six inches beyond the inner and outer faces of the blocks. Backfill material should be well graded sand-gravel mix (preferable types GW or SW) compacted to 90% to 95% relative compaction as it is being placed in layers. Care must be taken that the geogrids are not damaged and properly engage the joint between facing blocks, and are of the proper length for embedment beyond the wedge rupture plane.

About geogrids

Geogrid is the reinforcement material placed in layers within the backfill. Geogrids are produced by a number of manufactures, each offering a choice of several materials and tensile strengths. The specified geogrid is delivered to the job site in rolls, generally six to twelve feet wide, and are cut to lengths required by design. Most have bi-axial strength, with the higher strength along the rolled axis (perpendicular to the spool). The geogrid is cut to the design lengths in the field.

Each manufacturer offers several grades of geogrids, each with a different strength. Test procedures in accordance with ASTM or Geosynthetic Research Institute (GRI) procedures establish the ultimate tensile strength for each type of geogrid. The long-term design strength (LTDS) is derived from the ultimate tensile strength value and includes safety factors for long-term degradation, allowance for damage during construction, material imperfections, and other strength-affecting factors. A further safety factor is applied for design, generally 1.5. Therefore a Long Term Allowable Design Strength (LTADS) would be LTDS / 1.5.

To be effective in creating the en masse soil, the geogrid must be anchored at each end: into the soil beyond the backfill failure plane (described below) and anchorage into the facing block joint to resist the tension in the geogrid. To accomplish this, the geogrids are laid in the joint between blocks. Pullout resistance is provided both by the coefficient of friction at the joint plus whatever engagement means is used. The latter can be by pins through the geogrid interstices, folding over a lip in the block, or other means proprietary to each block vendor. To establish connection values each block type must be tested for each anticipated geogrid. A typical connection value might be displayed as:

Peak Connection Strength = 425 + 0.27 N, with a maximum of 1900 lbs.

Where 425 is the value (pounds) of the proprietary geogrid engagement to the block; 0.27 is the tangent of the block-geogrid-block friction angle; and N is the weight of the overlying blocks. The generally accepted factor of safety for connection strength is 1.5.

Another connection value is Serviceability Connection Strength. This is a tested value for failure when the geogrid is pulled to an elongation of ¾". Because this is a failure condition no further safety factor is needed.

The peak connection strength and ¾" serviceability connection strengths are available from the block vendors web site or literature. It is also available from www.icc-es.org which makes available evaluation reports from the various vendors (ICC Evaluation Service, Inc., Legacy Reports).

The factor of safety for peak connection strength is generally 1.5, and factors included should be possible damage during installation, material degradation, creep of the textile, manufacturer, and ratio of ultimate tension capacity and design tension.

Gather design criteria

After determining the site requirements such as retained heights along the length of the wall, plan view alignment and contouring that may require a sloped backfill, you will need characteristics of the natural (in-situ) soil both behind and below the reinforced zone. This will be the density of the soil and its angle of internal friction (phi value). This information will be provided by the geotechnical engineer, and will include other recommendations, such as the need for a global stability check if sloping conditions exist above the wall or underlying soil is questionable. Also needed is the density and phi value for the backfill material. Backfill material should be well drained sand/gravel mix, preferably Group GW on the Unified Soil Classification System. See Appendix B.

Determine retained heights, soil properties (densities and friction angles for both in situ ("external") and backfill ("internal") material, loads (surcharges and/or seismic design), and site space available.

NOTE: If slope conditions exist above or below the wall, consult the project geotechnical engineer to determine whether a global stability analysis is required, and if additional geogrids are needed to satisfy global stability requirements. When a slope exists at the base of the wall increased embedment is required. Typically, it is expressed as minimum embedment required for 5' to daylight at the base of the wall. This means that for a 2:1 slope, the embedment is 2.5', or for a 1:1 slope the embedment is 5'.

Select masonry units

Select the block vendor for texture, color, size and configuration desired. This is often dependent upon proximity to distributors.

Internal" and "External" forces

The term Internal Forces describes the lateral earth pressure within the soil-reinforced zone. This pressure applies force against the wall and creates the tension on the geogrids to maintain the integrity of the soil mass.

External Forces describe the lateral earth pressure acting outside and against the reinforced soil zone.

Determine lateral soil pressures

Both internal and external forces must be considered because the properties of the two soil likely will be different (density and friction angle), therefore the K_a factor needs to be computed for each soil. The Coulomb equation is generally used for both.

The Coulomb Equation is shown below, and note that because the resultant is assumed to act at an angle δ from the horizontal, the horizontal component must be computed. The vertical component is generally ignored.

$$K_a = \frac{\sin^2 (\alpha + \phi)}{\sin^2 \alpha \, \sin (\alpha - \delta) \left[1 + \sqrt{\dfrac{\sin (\phi + \delta) \, \sin (\phi - \beta)}{\sin (\alpha - \delta) \, \sin (\alpha + \beta)}} \right]^2}$$

$$K_a \text{ (horiz.)} = K_a \cos\delta$$

β = Angle of backfill slope
ϕ = Angle of internal friction of either backfill or in-situ soil.
α = Wall slope angle from horizontal (90° + batter angle from vertical)
δ = Angle of friction between soil and wall
(usually assumed to be $2/3\phi$ to $1/2/\phi$) but for the external force applied to the reinforced zone δ is assumed equal to ϕ.

Select geogrid

The geogrid manufacturer and type is selected based upon the tension resistance required. This is based upon the depth of the geogrid and the vertical tributary area between geogrid layers. The lowest geogrid is usually placed in the first block joint above the base, then every second or third joint, but generally not exceeding two feet apart. Spacing between layers may vary with design requirements, but to simplify instructions to the contractor equal spacing is often used.
When a trial spacing is selected, the internal earth pressure to each geogrid is calculated. This is the force which must persisted by anchoring geogrid both to the block facing and embedded into the backfill soil a sufficient distance beyond the failure plane.

The georgrid to block connection strength is block specific and this information is available from ICC Evaluation reports or directly from the geogrid vendor.

Tension in each layer of geogrid, T_a, increases with its depth in the backfill, and can be computed by the equation below:

$T_a = K_{ahi}\, z\, \gamma\, s$

where K_{ahi} = horiz component of K_a based upon K_a of internal (backfill) soil; γ = soil density; z = depth of soil above layer, and s = tributary height to the layer.

$s = [(\text{height to layer above}) - (\text{height to layer below})] / 2$

(Note: an exception is the tributary height above the uppermost layer and below the lowest layer)

This value, T_a, would be the horizontal tension on the or geogrid per longitudinal foot of wall, and from this value the geogrid type (LTADS) is selected.

Determine geogrid embedment

Check required geogrid length for extending beyond the failure plane. The geogrid must extend beyond this soil wedge failure plane a distance adequate to resist pullout. The resistance to pullout is provided by the friction between the soil and geogrid (both top and bottom surfaces used), the weight of the overlying soil, and the friction angle of the soil. Additionally, a reduction intereaction coefficient, C_i, is used which is dependent upon the particular geogrid and the surrounding soil. The C_i value usually varies from about 0.70 to 0.90.

The equation for the required length beyond the failure plan, neglecting DL, LL, and additional soil in a backfill slope, is:

$$L_e = \frac{FS\, K_{ahi}\, S}{2 \tan \phi_i\, C_i}$$

FS = Factor of safety (1.5 minimum)
K_{ahi} = Coefficient of active pressure (horizontal component, internal soil).
S = Tributary height between next higher and lower geogrids.
ϕ_i = Friction angle of internal soil
C_i = Soil/geogrid interaction coefficient

Note that the above equation is independent of the overlay depth z and γ because these factors cancels out in the equilibrium equation:

$$T_a = K_{ahi}\, z\, \gamma\, s = 2 L_e\, \gamma\, z \tan \phi_i$$

To include dead load, live load (not recommended) and additional soil over the backfill slope, use the complete equation:

$$L_e = \frac{F_s\, T_a}{2 \tan \phi_i\, C_i \left[\left[D + \gamma\, z + \tan \beta \left(\frac{H}{\tan \alpha} - H \tan \omega\right)\right]\right]}$$

Where D = dead load surcharge

The overall length of one geogrid as required is $L_a + L_e$, where L_a is the length within the soil wedge plus the wall thickness, and L_e is the length beyond the failure plane. NCMA recommends extending an additional one foot; AASHTO an additional three feet.

Note that this base width is only the minimum for geogrid embedment and additional width may be required for overturning and sliding resistance as discussed below.

Determine depth of reinforced soil (total base width)

The base width is defined as the depth of the reinforced soil (to the end of the geogrids) plus the wall thickness. Although this is initially estimated from 60% to 80% of the wall height, it must be checked.

The criterion is the failure plane angle which extends upward from the base of the wall and defi nes the limit beyond which the geogrid must extend for proper embedment. This angle, measured from horizontal, can either be the Rankine failure angle $(45° + \phi_i/2)$ or the more commonly used Coulomb failure plane angle, recommended by NCMA. This angle is:

$$\alpha = \phi + \tan^{-1}\left[\frac{-\tan(\phi-\beta)+\sqrt{\tan(\phi-\beta)\left[\tan(\phi-\beta)+\cot(\phi-\omega)\right]\left[1+\tan(\delta-\omega)\cot(\phi+\omega)\right]}}{1+\tan(\delta-\omega)\left[\tan(\phi-\beta)+\cot(\phi+\omega)\right]}\right]$$

α = Coulomb failure plane angle measured counter clockwise from horizontal.
ϕ = angle of internal friction of the internal (backfill) soil
β = backfill slope, if applicable
δ = friction angle at wall-soil interface (usually $2/3\phi$)
ω = wall batter measured counter-clockwise from horizontal.

The Coulomb line is steeper than the Rankine for most cases; hence it requires a lesser embedment length.

For a given failure angle α, the distance from outside face of wall to failure line intercept, L_a, for any height h_x is:

$$L_a = \frac{h_x}{\tan\alpha} - h_x \tan\omega + t$$

Where t = wall thickness

The quickest way to check the minimum required base width of geogrid is to check the uppermost layer of length $L_{a+}L_e$ then add the front batter of the wall to height h_x which is $h_x * \tan\omega$.

To determine the available embedment depth for any height h_x:

$$L_e \text{ (available)} = B - t - \frac{h_x}{\tan\alpha}$$

When a base width is determined based upon required geogrid embedments, it may not be great enough for the reinforced soil block to resist overturning, which will be checked below.

Check overturning

When considering overturning for an MSE wall the entire reinforced soil zone is considered one mass, therefore the overturning force is the lateral pressure against the end of the reinforced zone – the extremity of the base width. For overturning calculations the reinforced mass is assumed to have a vertical plane (even if the reinforced mass is considered trapezoidal). If there is a sloping backfill, the pressure is against the full height of the vertical plane – from base to intercept with the finished grade.

The Coulomb equation is used, with the density and phi values being those of the in-situ soil, and interface friction angle, δ, is assumed equal to ϕ.

If a surcharge is present it is to be included -- both dead load and live load (but not live load if seismic is included (see Seismic Design below).

For overturning, the earth pressure is assumed to act at one-third the total height of the vertical plane, and surcharges at one-half the total height.

Therefore, the total <u>overturning moment</u> is:

$$OTM = K_a(horiz) \, \gamma \, \phi * H^3 \, 1/6 + K_a(horiz) \, (DL + LL) \, H^2 \, \frac{1}{2}$$

To calculate resisting moment the weights used are the reinforced soil zone; weight of soil in sloping backfill, if applicable; weight of the wall facing blocks; and surcharges, if applicable.

Resisting moments are taken about the outer edge of the base of the wall. If the wall is battered, this will increase the moment-arm distances.

Therefore, the total resisting moment, RM, is:

$$RM = wH \, (0.5t + 0.5 \tan \omega \, H) + (B - t)\gamma \, H[0.5(B - t) + 0.5H \tan \omega] + (B - t)(D + L)$$
$$[0.5(B - t) + 0.5H \tan \omega] + 0.5\gamma(B - t)^2 \tan \beta$$

Where w = weight of wall, psf; γ = density of backfill soil; Φ = friction angle of backfill soil; B = total base width; t = wall thickness; β = backfill slope;

ω = wall batter angle from vertical; H = height of wall, ft;

D and L = dead load and live load.

The stability ratio (factor of safety) is: RM / OTM, which should be 1.5 or greater per NCMA.

Check sliding at lowest geogrid layer:

Compute lateral force same as above with $z = (H - h_x)$,

Where h_x = height to lowest layer

Resistance is provided by both soil friction at lowest layer plus to geogrid connection strength which is obtained from equation provided by block vendor, taking the form (1500 + 0.28N), in pounds, where N = weight of wall above, or:

Resistance = $W_{earth} \tan \phi_r + (XXXX + 0.XX \, N)$

ϕ_r = Friction angle of retained soil.

X = Values provided by vendor.

Check sliding at base:

For overturning/resisting, use the same driving force as above for overturning:

Sliding force = K_a(horiz) [γ ϕ_e H_1^2 0.5 + (DL) H_1]

Note that H_1 is the total height at the back of the reinforced zone from base to intercept with the sloped backfill surface. Therefore, $H_1 = H + (B - t) \tan \beta$. Sliding resistance is provided by friction between weight of reinforced soil mass plus the weight of the wall.

$W_{total} = W_{wall} + W_{earth} = wH + \gamma [(B - t)H + 0.5(B - t)^2 \tan \beta]$

Friction Resistance = $W_{total} \tan \phi_e$ C_i

Sliding Safety Factor = Friction Resistance / Sliding Force

Check soil bearing pressure:

For SRW walls the Meyerhof Method is used to determine bearing pressure. This assumes a rectangular pressure distribution under the footing, as opposed to a triangular distribution. The total vertical force is distributed uniformly over an effective base width. The effective base width is less than the full width by a distance equal to twice the eccentricity of the imposed load on the full footing width (easily verified with a diagram).

e = [(resisting moment) – (overturning moment)] / (total vertical load)

B_e = effective bearing width = $B - 2e$, where B is the total bearing width.

Bearing pressure = W_{total} / B_e

Soil bearing capacity:

Ultimate bearing capacity is calculated using the classical Terzaghi equation:

$Q_{ultimate} = \gamma d N_q + 0.5 \gamma B_e N_\gamma$ (an additional term to include cohesion is omitted because cohesion is usually assumed zero)

Where:

γ = density of underlying (in-situ) soil

d = depth of embedment of bottom block, ft.
B_e = effective bearing width, ft. (see above for methodology)

N_q and N_γ are non-dimensional coefficients per table below from the NCMA Manual. For these equations refer to Bowles' Foundation Design & Analysis, Fifth Edition, page 220 - 223.

ϕ_i	N_q	N_γ
30	18.4	22.4
31	20.63	26.0
32	23.2	30.2
33	26.1	35.2
34	29.4	41.1
35	33.3	48.0

| 36 | 37.8 | 56.3 |

Minimum safety factor for soil bearing per NCMA is 2.0

Seismic design – MSE walls

If seismic design is required for your locale or applicable code, the first step in the design is to determine the site specific seismic acceleration factor, k_h, which is a function of the Peak Ground Acceleration (PGA) for the site. The PGA can be determined from seismic hazard maps in IBC or from ASCE-7 '10. This can also be obtained by entering the longitude and latitude of the specific site. http://earthquake.usgs.gov/designmaps/us/application.php .

PGA can be computed as $S_{DS}/2.5$, or 2/3 of MCE where NCE is mead Peak Ground acceleration (PGAm) per ASCE 7-10.

NCMA recommends $k_h = A / 2$ for internal stability (Wall and reinforced zone) and $K_h = (A - 1.45) A$ for external stability (acting on reinforced zone). A = Peak Ground Acceleration.

In areas of high seismicity, however, the above can yield improbably high design accelerations. For example, for PGA = 0.40, $k_h = 0.42$ or external stability, results in unreasonably high seismic forces. Consequently, in these areas it is common practice among engineers to use a maximum value of $k_h = 0.15$ based upon slope stability analogy. Furthermore, NCMA states that "In practice, the final choice of k_h in any calculation may be based upon local experience, and/or prescribed by local building official or other regulations."

For insertion into the Mononobe-Okabe (Modified Coulomb) equation for K_{AE} you must convert k_h to an angle θ: $\theta = \tan^{-1} k_h$

If seismic design is required, three components must be considered for overturning and sliding stability:

1. Seismic inertial force from wall, F_1.

    The wall internal force is: $k_h w H$, where w = unit weight of wall in psf, and H = height of wall. This force acts at one-half the wall h.

2. Seismic inertial force from earth pressure within the reinforced zone, F_2.

    For this inertial force a depth of reinforced zone need not exceed one-half the height of the wall.

    Therefore $F_2 = k_h \gamma [(0.5H - t) H + 0.5 (0.5H - t)^2 \tan \beta]$

3. Seismic force acting on the back of the reinforced zone, F_3.

    This component is applied to a vertical plane at the back face of the reinforced zone, using a height increased by sloped backfill if applicable. The force may be reduced 50%.

    For this force use the Mononobe-Okabe (seismic modified Coulomb) equation below. The value, K_{AE} is for both static and seismic; therefore you will need to deduct K_A (static) to determine the increased force because of seismic, designated ΔK_{AE}

$$K_{AE} = \text{active earth pressure coefficient, static + seismic}$$

$$= \frac{\sin^2(\alpha + \theta - \phi)}{\cos\theta \sin^2\alpha \sin(\alpha + \theta + \delta)\left[1 + \sqrt{\dfrac{\sin(\phi + \delta)\sin(\phi - \theta - \beta)}{\sin(\alpha + \delta + \theta)\sin(\alpha + \beta)}}\right]^2}$$

Where $\theta = \tan^{-1} K_h$, α = wall slope clockwise from horizontal, (90° for a vertical face), ϕ = angle of internal friction, β = backfill slope, and δ = wall friction angle.

The horizontal component is $K_{AE} \cos\delta = K_{AEH}$

For this case $\alpha = 90°$ and $\delta = \phi_{extermal}$

Thus: $F_3 = (K_{AE} - K_a)\gamma H_1^2\, 0.5 + (K_{AE} - K_a)\, D\, H + k_h\, w\, H_1$

γ = density of soil, back fill or in-situ depending upon case. D = dead load surcharge. Note that the value H for this force is the wall height + added height because of sloped backfill, hence:

$H_1 = H + (B - t)\tan\beta$

These three seismic components must be added to static sliding and to increase overturning moment.

Increase in sliding force = $F_1 + F_2 + F_3$
Increased overturning = $F_1 (H/2) + F_2 (H + 0.5H \tan\beta)\, 0.5 + F_3\, 0.6\, H_1$

If seismic forces are included, the safety factor for sliding and overturning can be reduced to 1.1.

Added seismic tension to a layer is calculated by:

$K_h [(h_+ - h_-) / 2]\, w + \Delta K_{AEH} [(h_+ - h_-) / 2] [0.8 - 0.5 ([(h_+ - h_-) / H]$
where h_+ and h_- are heights of next higher and next lower layers.

Before starting with a seismic design for a SRW always check with the building department or agency having jurisdiction to verify applicability and any specific requirements.

Building codes & standards

IBC 2015 does not directly address segmental retaining walls.

The current standard design manual published by the National Concrete Masonry Association (NCMA – www.ncma.org), is:

Design Manual for Segmental Retaining Walls, 3rd. Edition (NCMA)

Acceptance Criteria for segmental retaining wall manufacturers is published by ICC Evaluation Services and can be obtained from www.icc-es.org.

Major SRW vendors also offer design handbooks plus other resources, most downloadable in pdf from their web sites.

Also see: Bowles' Foundation Analysis & Design, 5th. Edition, Chapter 12.

Getting help

In addition to the references above, all major SRW block vendors have web sites and offer technical support for their products. Some offer free software. Three major vendors are Keystone Retaining Wall Systems, Anchor Block Retaining Walls and Allan Block. Others can be found through a Google search for "segmental retaining walls". For design assistance, some firms specialize in MSE wall design.

TIP: Once Source:ABI Consultants www.abiconsultants.com*, Santa Ana, CA, 1-888-220-5596.*

Swimming pools are constructed in a wide variety of shapes, sizes, curvatures and designed to fit a specific terrain and soil conditions. One thing nearly all have in common is shotcrete or Gunite walls and bottoms sprayed over a shaped excavation, and encasing the reinforcing. Plaster or tile is used to provide a smooth, aesthetic finish.

The terms "shotcrete" and "Gunite" are used interchangeably, but the former refers to wet-mix spraying whereas the material is mixed in a hopper before exiting the nozzle, whereas the latter is a "dry-mix" where the material reaches the nozzle dry where water is injected. Shot Crete (we'll use the generic term) sticks to the earth and self compacts because of the velocity of application, thereby permitting it to be used against vertical surfaces. Shot Crete is covered in IBC '12, section 1910.

Designing the walls of a pool is unique because not only does the wall usually curve as it descends, but the strength of the cantilevered wall must resist the greater of earth pressure acting inward with the pool empty, or the water pressure outward if the exterior grade is lower or of poor soil. The design task is made further tedious because of the number of cross sections which must be checked (shallow end, deep end, and intermediate points).

The typical controlling condition is when the pool is empty and earth pressure from the outside governs the design. However, the condition is often reversed, such as for "infinity pools" or architectural features where the outside grade is substantially lower, or slopes downward lessening its lateral support value. There also may be lateral support from of a surrounding deck at or near the top of the wall. All these conditions must be considered and the walls of the pool designed for the most critical combination of conditions that may occur. Lateral loading from a surcharge or increased soil pressures because of expansive soil must also be considered.

Design of swimming pools is a specialty for some engineers and they have developed software (usually spreadsheets) to make the task less tedious.

The walls and bottom are generally at least 4″ thick, generally 5″ for floors, and may be more depending upon design requirements. Typically, #3 bars are used because of the relative ease in bending and securing to curved surfaces. Number 4 bars can also be used, but #5 bars are difficult to bend and place. Shot Crete strength is typically 2500 psi minimum, and a low slump suitable for pumping and spraying. Minimum reinforcing for flexural members is $200 / f_y$, $= 0.0033$ for Grade 60 reinforcing. Thus, for a 4″ wall the minimum would be #3 at 9″, however, the typical practice pattern is 12″ on center each way. Under slab drainage is recommended on sites with expansive soil with special reinforcement and/or thickened slab required for sites with expansive soil to protect from uplift along the bottom of the shallow end.

The classic method of designing swimming pool walls has been to draw to scale (or CAD generated) a cross section at each location to be investigated. Then divide the wall into segments, usually 12″ high. You can then determine the bending moment and shear at the bottom of each segment by constructing a table (spreadsheet) showing the active pressure from either earth or water acting at the bottom of each segment, and the additive (or deductive) moment due to the vertical weight of the segments above acting at their eccentricity from a reference point. This is illustrated in Figure 16.1. This is a tedious process but yields satisfactory results for design. Reinforcing is usually placed in the center of the wall, but for higher moments thicker walls may be needed and off-center reinforcing as the design may require.

The equation for moments at the bottom of any wall segment, pool full, then becomes:

$$M_y = (K_a \gamma h_y^3)/6 + (K_a w h_y^2)/2 - (W_1 e_1 + W_2 e_2 .....) - (62.4 h_y^3)/6$$

Where M_y = moment at depth y; K_a = Rankine or Coulomb active pressure coefficient; h_k = earth height above bottom of segment height; γ = soil density, pcf; w = surcharge in psf; W_x = weight of segment (50# for one foot high at 4" thick); e_x = eccentricity of segment x from reference point; and h_x = height of water above bottom of segment. This calculation is performed for each cross section and the critical condition (earth or water pressure controlling) determined. See Figure 17.1.

I once noticed the reinforcing on a neighbor's soon-to-be gunited wall was placed on the wrong side of the pool wall for his site conditions. His engineer was called and corrections made. Caveat: Always understand the client's possibly unique condition and don't rely on "standard" plans which may not fit the condition.

Caution: The most critical condition will control, such as when the pool is empty (no water pressure) or if the grade on the outside is lower such as for an "infinity" pool.

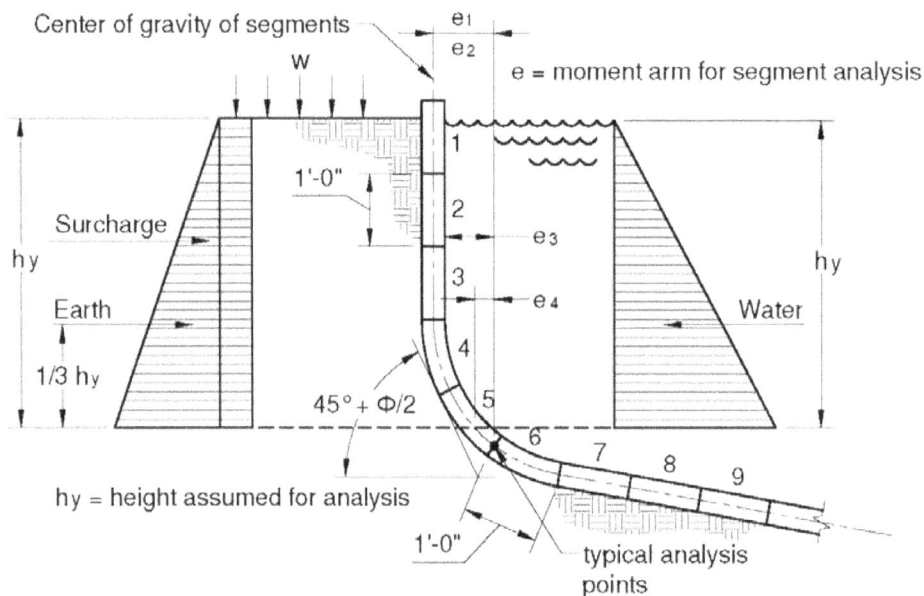

Figure 17.1 Analysis of swimming pool wall

Before attempting a pool design you may want to check for an engineer specializing in this type of work. Their expertise and software may be cost effective and assure proper design.

TIP: One source: Pool Engineering, Anaheim, CA, www.pooleng.com, 714-630-6100.

18. PILASTER MASONRY WALLS

Description

Shown in Figure 18.1 is a retaining wall with spaced pilasters and masonry filler walls. Such walls can be economical for low retaining or freestanding walls. The filler walls, usually 6″ or 8″ masonry, span horizontally between pilasters and the pilasters cantilever up from the footing.

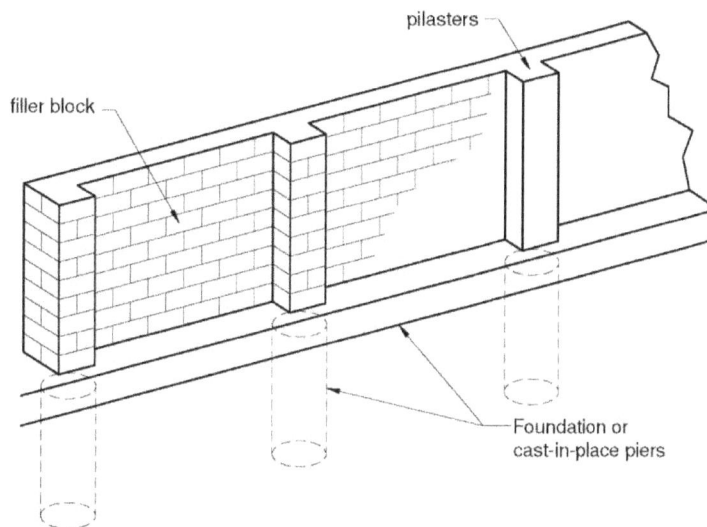

Figure 18.1 Pilaster masonry wall

Filler Wall Design

The filler wall spans horizontally between pilasters and those walls usually control the spacing of the pilasters. Freestanding walls are designed for wind and, if applicable, a seismic force. Horizontal reinforcing is placed in the center of the wall because lateral wind and seismic loads can be from either direction. To take advantage of continuity, it may be more economical to place the horizontal reinforcing at the center and design for the controlling positive (mid-span) or negative (at pilasters) moments, generally use $w(L)^2/12$.

If the filler wall retains earth, some or all of the courses will be subjected to lateral earth pressures and this controls the thickness of the filler wall. In that case, vertical reinforcing should be placed on the earth face between pilaster supports. Reduce reinforcing higher up the wall as moment decreases. The first step would be to determine the lateral pressure at the base of the wall, then select a wall thickness and vertical reinforcing, and then the reinforcing to span between pilasters.

A minimum amount of horizontal and vertical reinforcing should be used. The combined total area should be .0002bd, with not less than .0007 in either direction. Vertical reinforcing is often #4 bars at 32″ o.c. or 48″ o.c.

Pilaster Design

Pilasters are usually 16″ by 16″ masonry units, or smaller for lower walls and usually spaced 6″ to 8″ apart. Use conventional procedures for the design. Lateral load reaction to the pilasters from the filler walls will be triangular if retaining earth, and uniform if wind-only or seismic loads. Reinforcing usually consists of four bars and lateral ties.

Alternatively, design the core as reinforced concrete, specifying high-strength concrete (3,000 psi or greater). This results in nearly the same moment capacity as the full CMU pilaster block using the same reinforcing.

Footing Design

Only a nominal footing is needed under a filler wall. Pilaster footings can be either conventional rectangular spread footings, or cast-in-place piers in drilled holes.

If pilasters are cantilevered from an embedded pier, if not constrained at the surface, the point of contra flexure for moment is below the ground surface. This is often assumed to be one-third the embedment depth. Some engineers and tests suggest a more realistic point of contra flexure is 1/5 to 1/6 below ground surface.

For the design of drilled piers see Chapter 8 – Pile and Pier Foundations.

Description

Retaining walls are broadly defined as either yielding or non-yielding. The former refers to cantilevered walls, which are free to rotate, thereby allowing a lateral displacement at the top which activates the soil wedge concept upon which both Rankine and Coulomb theories are based.

Non-yielding walls are restrained at the top to prevent movement and therefore generate a reaction at the top and reduce moments at the base of the wall. A typical restrained, non-yielding, wall is the so called "basement wall". The designer must assess whether the wall really is "restrained" at the top against lateral movement. Wood diaphragms may be too flexible to provide sufficient restraint to allow a K_o condition to develop.

Lateral restraint at the top may be accomplished using tie-backs, also called anchored walls. These walls use drilled and grouted anchors placed into the backfill as the wall is constructed to provide a reduced lateral force on the wall, but may not achieve a K_o condition. If multiple levels of lateral restraint are required, such as for a multi-level structure, the design becomes complex because of varying wall moments, shears and reactions. Tie-back forces can also be affected by earth movement. Be careful here. If the tie-backs yield even slightly, the K_o condition may not be achieved.

Dual Function Walls

Often it is desirable to prepare two designs for the same wall. For example a basement wall may be backfilled before the lateral restraint at the top (such as a very rigid floor or roof diaphragm) is in place. It can first be designed as a conventional cantilever wall as for an assumed depth of backfill, and perhaps lessening the factors of safety because of a temporary condition. This would require a larger footing for overturning and result in a larger moment at the stem base. Then a second design is prepared for the final condition when the top restraint is in place and backfill completed. Then you've covered both conditions, but only if the contractor placed the backfill to meet your design/soil placement assumption.

If the bottom of a basement wall is fixed at the footing at the top, and assuming a triangular earth pressure against the wall, the base moment at the bottom of the wall will be about one-half the pin-pin positive moment of that for a cantilevered wall, and the positive wall moment will be about one-quarter the positive moment condition at the base of the cantilevered wall.

"At Rest" Active Soil Pressure

If a wall is restrained from movement at the top and therefore the sliding-wedge active pressure cannot be mobilized, the lateral soil pressure is somewhat higher. This is termed the "at rest" pressure, (designated K_o) and is applicable to a wall rigidly restrained at the top, such as a basement wall (but light framing with a flexible diaphragm may be inadequate "restraint" and the active soil wedge may be activated). The at-rest soil pressure for a level backfill is: $K_o = 1 - \sin \phi$, where ϕ is the angle of internal friction. For example, if $\phi = 34°$, $K_a = 0.44$, as opposed to $K_a = 0.28$ (assuming level backfill). For sloping backfill, a suggested equation is

$$K_o = (1 - \sin \phi) \left(\frac{1}{x} + \sin \beta\right).$$

For a well-drained granular soil, with a level backfill a typical value for $K_o = 0.50$. For a submerged sandy soil the density could be 125 pcf giving a lateral pressure of 0.5 (125 – 62.5) + 62.4 = 93.7 pcf. Clayey soil can be higher. Some agencies require $K_o = 1.0$, giving 110 pcf for a soil density of 110 pcf. ASCE 7-16 specifies a minimum of 60 pcf for "relatively rigid" walls, and states that basement walls not more than 8 feet below grade and with light roof framing (flexible) are not considered "rigid". Lateral pressure diagrams for restrained walls are shown in Figure 18.1. You are advised to get design values from the geotechnical engineer and check applicable code requirements.

An alternate to the triangular lateral pressure distribution which, some geotechnical engineers specify is a uniform pressure as shown in Figure 18.1 (b) or (c). This pressure diagram is usually used for open-end excavation and may not be applicable for backfilled restrained walls where a triangular distribution assumption would be more appropriate. Note that the clipped top and bottom corners in (b) can be ignored – a full-height uniform load will give only slightly more conservative wall moments. This uniform pressure, for sandy soil, is often defined as: $0.65 \gamma H \tan^2 (45 – \phi/2)$. Given a level backfill this corresponds to $0.65 \gamma H K_a$. This method results in about 25% higher wall moment than an equivalent triangular pressure using the same K_o. See Figure 18.1.

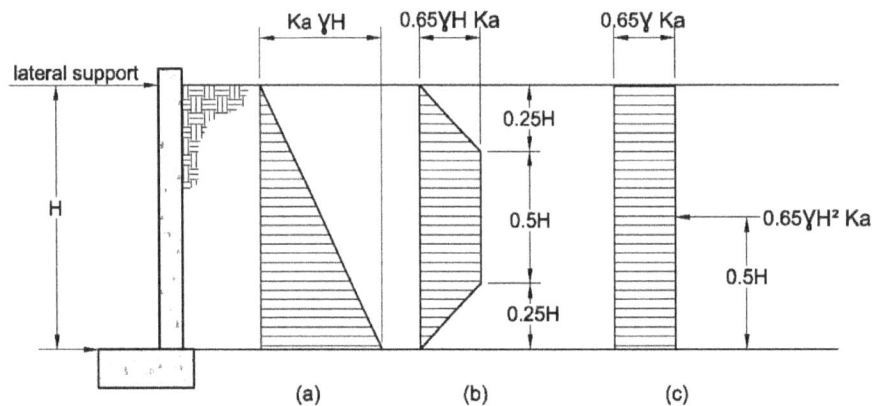

Figure 19.1 Lateral pressures diagrams for non-cohesive soil

Seismic Force on Non-Yielding (Restrained) Walls

Several Texts (e.g.Kramer) proposed the following formula (slightly revised):

$\Delta P_{eq} = \gamma \, k_h \, H^2$, acting at a resultant height of about 0.6H

Where ΔP_{eq} is the added lateral seismic force, γ is the unit weight of soil, and H is the retained height.

k_h = the horizontal seismic acceleration factor as used in the Monobe – Ohabe equations.

The resultant acting at 0.6H gives a slightly trapezoidal force diagram, however, for ease of calculation a uniform load can be assumed with less than 2% conservative error for position moment.

It should be noted that there are so few incidents of earthquake damage to such walls so that many experts agree that seismic design of restrained (e.g. "basement") walls may not be necessary, particularly given an adequate factor of safety for the service level design.

Description

Sheet piles are driven into the ground to retain earth while excavation is done on the opposite side. Sheet piles can also be included in permanent retaining structures alongside waterways (bulkheads and quay walls). Most sheet piles are steel, configured in an interlocking Z-shape to increase bending capacity, and to resist stresses during driving. Pre-stressed concrete panels are also used for sheet piling.

Sheet piles derive lateral support from embedment into the soil below the base grade, and can either cantilever up from that level, or be laterally restrained near the top by tie-backs in which case a horizontal member must be provided spanning between tiebacks.

Design Procedure

The design considerations for sheet piling involve the following:

1. The embedment into the base soil must be adequate to resist the total active lateral thrust by passive pressure in the embedded depth.

2. The bending capacity of the sheet pile material must be checked at the point of maximum moment. The point of contra flexure (zero shear and maximum moment) is usually about one-third down the embedded depth (although some texts state the actual point of contra flexure is closer to 1/6 the embedment depth), which increases the design moment. Manufactured sheet piles are designed to resist the driving impact during pile driving.

3. Tie-backs, if used, must extend beyond the line of soil rupture a sufficient distance to mobilize adequate passive pressure resistance for the anchoring device. Using tiebacks will reduce pile size, depth of embedment or required section modular of the pile.

4. The design of sheet piling is based upon the soil design parameters recommended by the geotechnical engineer. Input from the sheet pile vendor (most have handbooks and some have software to assist) and involving an experienced sub-contractor are essential.

Waterfront structures must consider impact from docking ships. (Incidentally, in marine work, the outboard bottom soil level, below the water line, is referred to as the "dredge line".)

A generalized force diagram of a sheet pile wall is shown in Figure 20.1. Note that by statics the horizontal active pressures and passive resistance must balance. Determining "d" and "D" shown in the illustration can be determined by statics and is an iterative (trial and error) process.

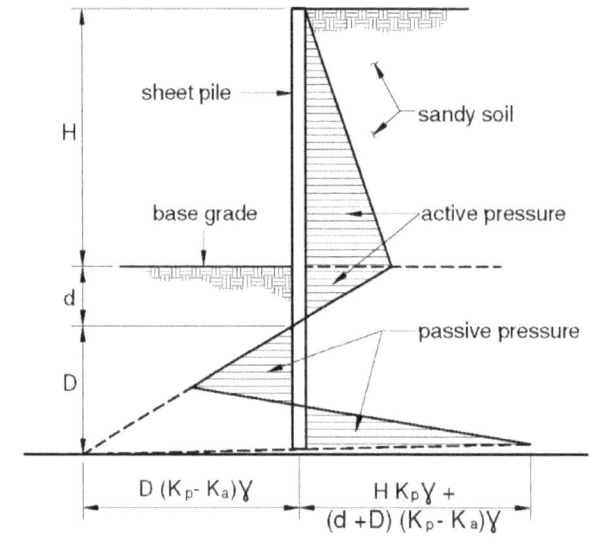

Figure 20.1 Generalized cantilevered sheet pile pressure diagram – sandy soil

References

The design of sheet pile walls is complex and references should be consulted. One good reference is Das' Principles of *Foundation Engineering, 5ᵗʰ Edition*, Chapter 9. Another: Rayapakse *Pile Design and Construction Guide,* 2003. Teng's *Foundation Design*, Chapter 12, is very good with tables and examples. Contractors specializing in sheet pile installation are the best source for economical design and site-appropriate recommendations and vendors of sheet piling have essential design data. A Google search for "sheet pile design" will yield valuable sources.

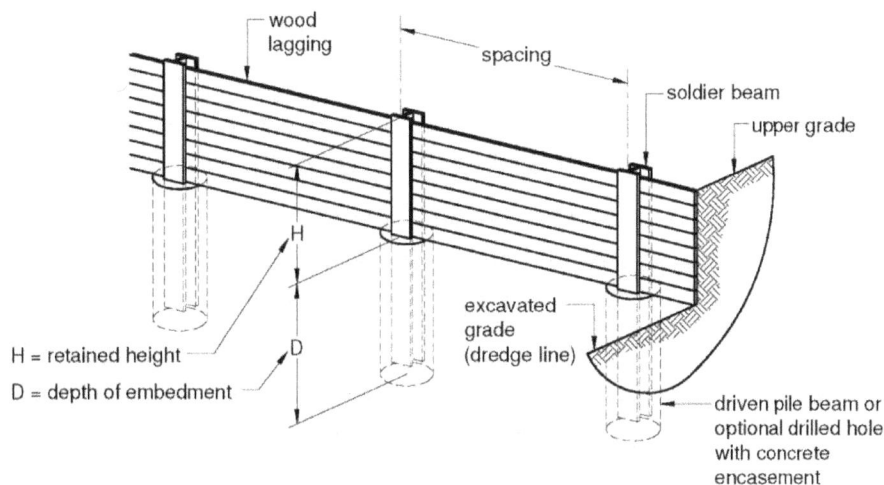

Figure 21.1 Typical soldier beam construction

Description

Soldier beam retaining walls are used to temporarily retain soil, such as at a construction site. Such retaining walls can also serve as permanent retaining walls as shown in Figure 20.3. This concept is illustrated in Figure 21.1. Steel HP (wide flange) beams are driven into the soil a sufficient embedment depth to resist by passive pressure the lateral force and moment imposed by the retained soil. The soldier beams (also called soldier piles) are usually spaced six to eight feet apart and can also be dropped into pre-drilled holes and encased in lean concrete. Soldier beams are usually cantilevered, but if space is available, and for retained heights over about 15 feet, tiebacks can be used to reduce the beam size and depth of embedment.

As excavation proceeds on the down-grade side, wood lagging is placed horizontally to support the retained soil. Lagging is supported at each end by the beam outer flanges.

Design Procedures

Consult with the geotechnical engineer for design criteria. This information will include nature of the soil, phi angle, soil density, active and passive allowable pressures, arching factors and any other site-specific recommendations. It is advisable to also consult with the contractor to verify the availability of the most economical beam selection and any other concerns he or she may have.

There are numerous design methodologies used and most foundation engineering textbooks propose various design approaches. This text selected a relatively simple procedure which is often used. This procedure assumes non-cohesive (sandy) soil. If the soil is clay a different passive resistance diagram will apply, if cohesion is incorporated into the design, and the geotechnical engineer should be consulted. It should be noted that although clay is usually assumed to have a zero phi angle, it actually can vary in a range as low as 6° to 12° or more.

Whether to use tiebacks is another decision to be made. The following procedure assumes a cantilevered system in sandy soil.

A basic design requires these steps:

There are many theories presented for calculating embedment depths (Plum method and others) and that shown in Figure 21.2 is one that closely agrees with other methods.

Determine the driving forces, that is, forces imposed by any construction surcharges and the active soil pressure tributary to each pile beam. Use the Rankine equation to calculate K_a. Several designs may be done to optimize the beam spacing based upon lagging selection, embedment depths, and beam sizes.

Referring to Figure 21.2, after P_a and P_w have been calculated, the depth of embedment must be determined. This will be a function of the allowable passive pressure and arching factor allowed to increase the effective flange width, or hole diameter if pre-drilling is used. The arching factor, A, can be taken as 0.08 * phi, but should not exceed about 2.5. This means that the effective pressure width in front of a 24″ diameter drilled and concrete filled beam encasement, with a phi of 32° would be 0.08 * 32 = 2.56, but use 2.5. Thus the effective passive pressure would be a width of 2.0*2.5 = 5.00 feet which will considerably reduce embedment depth and moment applied to the beam.

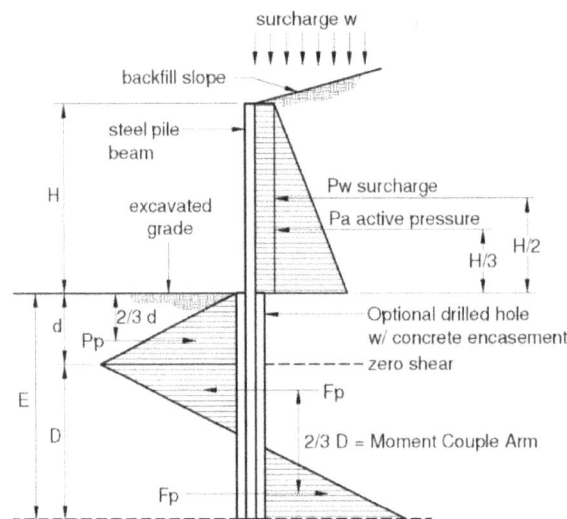

Figure 21.2 Theorized forces on cantilevered soldier beam in sandy soil

For determining the embedment depth to zero shear (where beam bending is maximum), designated "d", the following equation can be used:

$$d = \sqrt{\frac{P_P \times SF \times 2.}{p \times Dia \times A}}$$

Where P_p is equal to and counteracting to $P_w + P_a$; SF is the safety factor applied to allowable passive pressure; A is the arching factor multiplier; Dia is the hole diameter or flange width, whichever applicable; and p is the allowable passive pressure in pcf at "d". Consult with the geotechnical engineer to determine if the recommended passive pressure includes a factor of safety; if it does not, you could let SF=1.5.

1. The maximum beam moment is then determined by summing moments above the point of zero shear. Mathematically the result is equivalent to:

$$M_{max} = P_w (0.50H + 0.67d) + P_a (0.33H + 0.67d)$$

2. The maximum moment is resisted by a passive pressure couple consisting of $F_p * 0.67D$. Therefore the required depth D can be estimated from the following equation:

$$D = \sqrt{\frac{(\text{max. moment}) \times SF}{(p \times Dia \times A \times d \times 0.25)0.67}}$$

    The required depth of embedment is then (d + D). As a rule-of-thumb for sandy soils this is usually in the range of 1.3 H to 1.5 H. It is conservative to add 20% - 30% depth to calculated embedment.

3. After the maximum moment has been computed, convert it to LRFD (Load Resistance Factor Design) by multiplying by the usually applicable load factor of 1.6. Then select several beam options from AISC 13th edition, LRFD, Steel Design Handbook. When several beam selections are made it is recommended that you talk with the contractor for his opinion on which is most economical or available.

4. Select the lagging. Treated lumber should be used, and a conservative fiber stress is 900 psi. Calculate the lateral pressure at various depths, H_y,(to determine changing lagging thicknesses) which is $K_a * \gamma * H_y$. When the simple span moment is calculated it is acceptable to multiply by 0.8 because of arching action of the soil between pile beams. Lagging is either 3″ or 4″ by 12″ treated wood. Their ends should bear against the beam flange a minimum of 3″. Allow about 1″ between each lagging for drainage.

For a typical soldier beam wall construction (see Figure 21.3.)

Figure 21.3 Completed permanent soldier beam wall

The soldier beam wall shown in Figure 21.3 retains about seven feet and uses W14 wide-flange beams spaced eight feet apart embedded about eleven feet, and placed in drilled holes and encased in lean concrete. Lagging was 6″ x 12″ #1 Douglas fir.

If the retained height is over about 15 feet, and space is available behind the wall, tiebacks can be considered. Tie-backs will reduce embedment depth and beam size.

Usually tiebacks are steel rods inserted into small diameter holes drilled into the backfill a sufficient distance beyond the failure plane to provide anchorage after grouting, and are inclined downward at an angle of 15° to improve withdrawal resistance and to facilitate grouting. The ends of the tie-backs are welded to the steel beams.

The depth of embedment d is determined by taking moments about the tie-back height. Knowing, $P_P = (P_w + P_a) - $ (tie-back capacity), solve for "d". The horizontal component of the tie-back force is easily determined by statics. Consult with the geotechnical engineer or the installation contractor to determine the allowable perimeter shear stress between the grout and the project soil type to determine the grouted depth of embedment of the tie-rod beyond the failure plane.

A simplified force diagram is shown in Figure 21.4 and an alternate tie-back anchorage is shown in Figure 21.5.

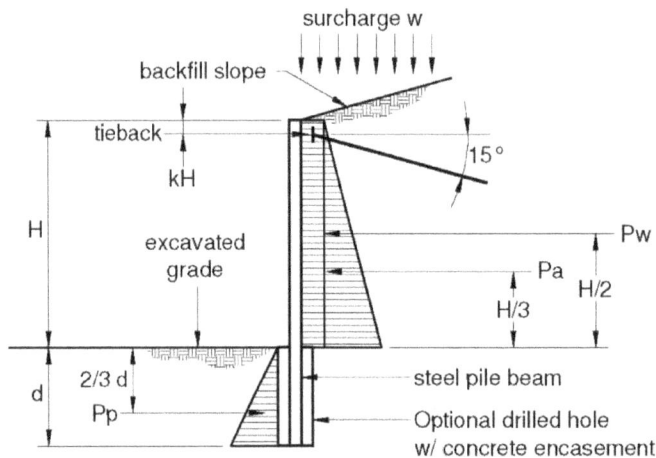

Figure 21.4 Tie-back force diagram

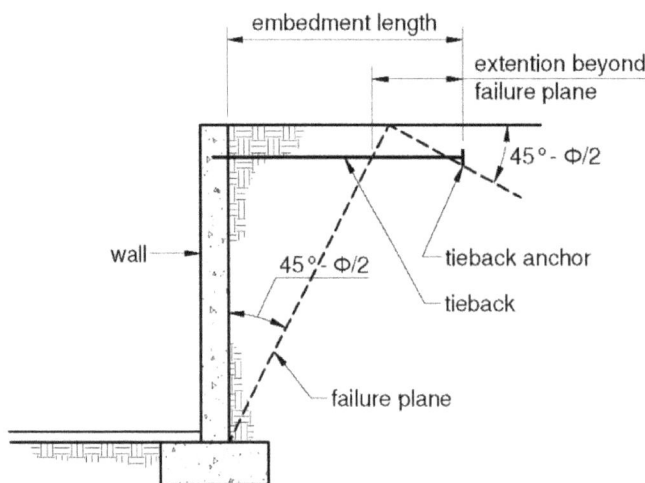

Figure 21.5 Tie-back anchorage

The above photo is a rare occurrence. Building permit not obtained, wall not engineered, minimal reinforcing in ungrouted cells, and other oversights.

"Failure" of a retaining wall does not necessarily mean total collapse as shown above, but rather local signs of impending instability and likelihood of a total collapse. Total collapses are relatively rare. In a total collapse the wall overturns, slides, topples, or otherwise causes a massive letting loose of the retained earth with resulting damage above and below the wall. Such walls cannot be saved – the remedy is rebuilding. The engineer who provided this photo was retained to investigate the deficiencies causing the collapse and to design a new wall.

Fortunately, retaining walls are quite forgiving; nearly always display telltale signs of trouble alerting an observer to seek professional evaluation before a collapse. After an evaluation, and determination of the causes, most walls can be saved.

The most common sign of distress is excessive deflection of the wall – tilting out of plumb – caused by a structural overstress and/or a foundation problem. Some structural deflection is to be expected and a rule-of-thumb is $1/16^{th}$ inch for each foot of height, which is equivalent to one-half inch out-of-plumb for an eight foot high wall. More than that is suspect. It's easy to check with a plumb bob.

Here are Twelve Things That Can Go Wrong and Signal Distress:

> 1. <u>Reinforcing not in the correct position</u>. If the stem shows sign of trouble (excessive deflection and/or cracking) the size, depth, and spacing of the reinforcing should be verified. Testing laboratories have the devices (usually a magnetic field measuring Pachometer) which can locate reinforcing and depth with reasonable accuracy, up to about 4 inches depth. For exact verification you can first locate the reinforcing then chip out the concrete, blocks, or core the wall to determine its exact depth and bar size. More elaborate devices are also available if needed – check with a testing laboratory, they will come to the jobsite. Unbelievably, cases have occurred where the reinforcing was placed on the wrong side of the wall, either through a detailing error, or contractor error. When the actual reinforcing size, location, and spacing are determined, and perhaps a core taken of only the concrete to verify strength of the stem material, a design can be worked backwards to determine actual design capacity and thereby guide remedial measures.

2. <u>Saturated backfill</u>. Retaining walls are generally designed assuming a well drained granular backfill. If surface drainage is allowed to penetrate and accumulate in the backfill, the lateral pressure against the wall can double. Ponding of water behind the wall not only indicates poor grading, but clayey soil impeding the downward seepage of water. The surface of the backfill should be graded to direct water away from the wall, or by the use of drainage channels adjacent to the wall to intercept surface water and divert it to disposal. Often surface water problems are those attributable to a misdirected or poorly timed irrigation system. Poor backfill material, such as that containing clay, can swell and increase wall pressure. Most contractors always uses crushed rock for backfill; it's less expensive than pea gravel, and the elimination of tamping compaction of granular soil offsets the cost of crushed rock versus the use of materials that require compaction, and assures good drainage. Don't compact backfill by flooding.

3. <u>Weep holes that don't weep</u>. The only thing that comes out of most weep holes is weeds – not water. Weep holes become clogged when there is not any filtering, such as a line of gravel or crushed rock placed along the base to provide a channel for water to drain to weep holes, or to be conducted by an embedded perforated pipe. Commercial filtering fabric is available. Weep holes in masonry are usually made by omitting mortar at the side joints of every other block (32 inches on center). For concrete walls, 3″ diameter pipe sleeves are often used, spaced 4′ – 6′ on center, or as deemed appropriate by the designer. Specifying proper drainage measures (backfill material, surface water control, and base-of-wall drainage) is an important specification task for the EOR (Engineer of Record).

4. <u>Design error because of misinformation</u> Design errors as the cause of failures are relatively rare when prepared by an experienced designer. However, sometimes the designer is given insufficient or erroneous information, for example, "Design the wall to retain eight feet", but a later examination of the grading plans, or as-built conditions, shows the wall retaining nine feet, an additional foot, thereby increasing the base moment on the stem by 42%.

5. <u>Calculation errors</u>. An experienced designer can quickly spot a calculation error because it obviously "doesn't look right". New engineers usually lack that experience, and in such cases don't let the design leave the office without a check. A 15-minute review could save costly fixes and damage client relations. And don't assume a plan checker will find errors.

6. <u>Unanticipated loads</u>. Again, this is a client-to-designer information problem. Good communication is essential. Is there a surcharge the designer didn't know about? A steeper backfill slope? A beam connection? Wind load? A foundation investigation or memo that wasn't transmitted?

7. <u>Mistakes in using software</u> If software is used as a design aid, it is essential that the designer correctly inputs data and understands the capabilities and limitations of the particular program (Retain Pro advises its users to be licensed civil or structural engineers, or at least have the expertise to design a relatively complex retaining wall by hand calculations). If in doubt of a result, do a quick hand calculation.

8. <u>Detailing errors</u>. The contractor must have clear instructions. Details not conforming to the design, or doubtful of interpretation, must be avoided. Perhaps the biggest source of problems is with reinforcing placement. The senior author recalls a case where the designer actually detailed the rebar on the wrong side of the wall! In another case vagueness of details led to dowels from the footing extending only 6″ into the stem, rather than the intended 24″, because of confusing dimensions. Easy-to-read drawings and careful checking by the designer can eliminate these problems.

9. <u>Foundation problems</u>. When a soil investigation report is provided, there will be guidelines for design (allowable soil bearing, friction factors, seismic design factor if applicable) and any caveats based upon site conditions, such as liquefaction potential or recompaction of the underlying soil. Implementation of such recommendations should provide a trouble-free foundation. However, often such an investigation is not provided, calling for special care by the designer. Without such a report the soil bearing is limited by Code, usually to 1,500 psf, and the coefficient of sliding friction to 0.25, and allowable passive pressure to 150 pcf. Regardless of using more conservative values, the designer should make every effort to become aware of any adverse conditions, such as fill material, compressible soil, water table, or other factors that could cause excessive settlement – or reduce sliding resistance.

10. <u>Inadequate specifications and notes.</u> If you use "boiler plate" notes or specifications, edit those carefully and use a checklist. It's embarrassing to have notes or paragraphs that obviously apply to another project. Here's a note that should keep you out of trouble and avoid problems: "If a discrepancy arises between the drawings and field conditions, or where a detail is doubtful of interpretation or an unanticipated field condition is encountered, the engineer shall be immediately contacted for procedure to be followed. Such instructions shall be confirmed in writing and distributed to all affected parties". And another good one: "Wherever there is a conflict between details and specifications, or between details, or where doubtful of interpretation, the most restrictive shall govern as determined by the Engineer of Record."

11. <u>Shoddy construction</u>. This could be anything from a homeowner having built a wall from a "how to" book, to an inexperienced or unscrupulous contractor building without plans or not following the plans – inadequate grouting or mortar or improperly placing reinforcing. Retaining walls are quite forgiving and poor construction may not appear as distress for years, or never. An engineer tells the following: "I once built a vacation home with a five foot masonry retaining wall intended to be restrained at the top by the floor slab placed on the backfill. I instructed the contractor not to place any fill until the wall was properly braced since there was only a 14 inch wide footing. I went to the site a week later and to my amazement he had placed the backfill the full five feet – no floor slab yet – and the wall was perfectly plumb! Another case of practice defying theory – but don't count on it! Always, when possible, have a pre-construction meeting between the designer and contractor to be sure all conditions and requirements are understood, and jointly review the plans."

12. <u>Age</u>. If a retaining wall has been in place twenty years and does not show signs of distress, the chances are it will remain so for another twenty years, or fifty years. The adage "if it ain't broke, don't fix it" may be prudent advice. However, the caveat is that this precludes any change affecting the wall, such as new surcharges or a change in drainage above the wall. If in a seismic area the chances are it has already successfully withstood several earthquakes, but if the seismic risk is high and its failure could impact another structure, a seismic evaluation would be appropriate,

And Twelve Fixes That Could Save a Wall:

Note that each of the fixes listed below have been successfully used, but it is assumed that the wall is not in such distress that none are viable solutions.

1. <u>Correct surface drainage problems</u>. You can't economically replace the backfill or get to the base-of-wall drainage system, but you can re-grade at the surface so water does not collect behind the wall. Perhaps a small concrete diversion culvert. Often just shutting off an over active irrigation system will mitigate the problem. Additional weep holes can be cored through the wall, although perhaps visually objectionable.

2. <u>Reduce the retained height.</u> If the soil lateral pressure needs to be reduced, investigate whether re-grading of the surface can reduce the height of earth retained. Sometimes a change in landscaping, or a depressed drainage culvert at the back of the wall may reduce the height to an acceptable level based upon the as-built capabilities.

3. <u>Use tie-backs</u>. If the stem is severely overstressed an option is to use tie-backs extending back beyond the failure plane. Drill holes through the wall and install conventional tiebacks (also called soil nailing). A downside of this is the appearance of the tie-back anchors on the exposed face of the wall. Or perhaps a tie-back at the top of the wall can be used, with a concrete anchor block, or an added slab-on-grade. Using tie-backs requires re-analyzing the wall moments and shears because of the changed restraints.

4. <u>Extend the footing</u>. You can extend the toe of the footing and thereby substantially reduce the toe soil bearing pressure. Determine how much you need to extend the footing, then excavate to the bottom of the footing (excavate deeper for a key if necessary) and place concrete. To transfer shear and moment at the interface, drill holes in the existing footing and epoxy dowels to resist the calculated pullout induced by bending of the toe. It may be prudent to maintain lateral stability by excavating in front of the toe in longitudinal increments, say twenty feet, or less.

5. <u>Remove and replace backfill material</u>. This may be the only solution if saturated backfill is the problem and cannot be controlled at the surface. Use crushed rock backfill, and be sure the base-of-wall drainage is functional.

6. <u>Reinforce the front face of the wall</u>. This can be done by forming or pneumatically placing concrete to thicken the stem at the base, and tapering to a height where the added strength is no longer needed. This is on the compression side so the only design concern (other than how much thickness to add) is shear transfer at the interface, which can be accomplished by drilled dowel pins. This assumes, of course, that the existing footing will still be adequate.

7. <u>Add a key</u>. If there is a sliding problem you could add a deepened key in front of the existing footing. This will increase passive resistance and may be all you need. See #4 above for incremental excavating during this process.

8. <u>Use cantilevered soldier beams</u>. Drill holes on the heel side of the footing and embed a vertical beam, tied to the wall to transfer load to the beam. Space the beams at a distance the wall will span horizontally. The projection of the footing heel will determine how close to the wall the soldier beams (micro-piles?) can be placed.

9. <u>Get a building permit</u>. Quite often there is not any apparent distress in a wall, but an observant building inspector discovers that a permit had not been issued. This usually happens when a new building or addition is being constructed on the property. If plans for the wall are found it requires only substantiating calculations with an engineer's signature. If it can't be justified, then one of the procedures above may be needed to remedy an overstress. If plans cannot be found it's necessary to determine how the wall was built. This means probing and perhaps testing to determine location and spacing of reinforcing, toe and heel dimensions of the footing, and perhaps core tests of the wall material. The task then is one of working backward to find the capacity of the wall and hence its adequacy. Lesson: Always advise your client to get a permit; it could save a future expense.

10. <u>Push it back to plumb</u>. Not recommended, but has been successfully done in some cases if the wall is only out of plumb an inch or two, and not all backfill has been placed, and depending upon its height, and in conjunction with the above fixes, the wall can be pushed back to near-plumb. The wall may have been bumped to cause this, or tractor compacting too close to the wall. This is an arguable procedure but has been done successfully with few after effects. Use this with extreme caution! You may want to remove some backfill first.

11. "<u>Tear down that wall</u>". If it's in bad shape, and none of the above makes sense, it can be less costly to tear it down and rebuild. Especially valid if new conditions exist, such as need for a higher wall or a preference for a different wall material.

12. <u>An exotic solution?</u> We engineers pride ourselves on innovation. There may be a unique site condition that suggests a cost-effective fix. And you could come up with an ingenious method of saving a wall from reconstruction, and be a hero to a very happy client!

Some Actual Cases

Here are a few examples (edited) of problems that have occurred:

Case A. A wall was observed to lean excessively and it was found that the reinforcing protruding from the foundation was on the wrong side of the wall. Solution: Add tie-backs.

Case B. A wall was observed to lean excessively. Investigation revealed the wall had been designed to retain 12 feet of earth, with an extension of the wall another four feet above grade for screening. The owner and his landscaper arbitrarily added two additional feet of earth, thereby increasing the moment at the base of the stem by 60%! Solution: Add tie-backs.

Case C. Again the sign of a problem was leaning of the wall. Investigation discovered that the contractor had misinterpreted the plans and halved the number of dowels projecting from the footing. Solution: Gunite added wall thickness at the base, bonded to the existing wall.

Most of us have had similar experiences and other solutions.

Vertical Control Joints

Vertical joints along the length of the wall are intended to control cracking locations and are largely a matter of judgment. Shrinkage in a wall cannot be eliminated. As the adage goes, concrete shrinks and ice cream melts, or "if it ain't cracked, it ain't concrete." We can attempt to control where the cracks form by forming crack control joints and by increasing the horizontal reinforcing. With a little more than minimum reinforcing there are few reports of problems when control joints are 100 feet or more for masonry, and somewhat less for concrete. The more horizontal reinforcing, the less likely vertical cracks will be obvious, and the further apart joints may be spaced. In the case of a concrete wall, a ratio of $0.002A_{gross}$ is suggested; for masonry $0.0013A_{gross}$ is suggested (#5 bars at 32" o.c. for an 8" CMU wall).

Vertical joints for both concrete and masonry should be "cold joints", allowing for movement, but it is suggested that some horizontal dowels extend into the adjacent wall to assure out-of-plane alignment. Usually one end of horizontal dowels are wrapped, sleeved, or greased to prevent bonding.

Drainage

Improper drainage causing water seepage into the backfill is the leading cause of retaining wall problems. Lateral earth pressure design is usually based upon drained soil. Saturated soil can substantially increase lateral pressures. Therefore, it is important to have weep holes at the base of the wall for any percolating water to escape. In concrete walls drain holes are 3" to 4" in diameter to facilitate cleaning and spaced five or six feet on center. Gravel should be placed along the inside base for any water to freely flow, otherwise only weeds will be coming out of a weep hole.

"Weep holes" in masonry walls can be provided by leaving the head joints open at alternate blocks (no mortar in end joints at 32" on center).

In lieu of weep holes, or for basement type walls, horizontally placed perforated Sch 40 pipe should be laid along the base of the heel adjacent to the stem, slopped to an outlet, and encased in a generous amount of coarse gravel. It is also recommended to lay a filter fabric over the gravel to keep out soil fines.

The most important drainage control is to keep water off the top slope as much as possible. This can be done by slope control, paved swales, paving, or other means. <u>Preventing water from entering the backfill is critical important because it changes the soil characteristics and increases lateral pressures.</u>

Backfill

Backfill material should be sandy non-cohesive material. Clayey soil are to be avoided because clay swells when wet, causing additional lateral pressure. An excellent practice is to fill the soil wedge with gravel.

Compaction

Compact the gravel behind the wall with care. You don't want settlement to occur later. Place the gravel in layers about one foot thick and start compacting at the face of the wall and work away from the wall. Gravel is best compacted with multiple passes of a vibrating plate compactor.

Inspections

If a geo-consultant was employed, he or she will verify that the footings are excavated into the anticipated soil and indicate any corrections deemed necessary. They can also approve the backfill material.

Placement of reinforcing dowels projecting from the footing into the wall are critical to the design, and the Engineer-of-Record (EOR), or a deputy inspector, should verify that the dowels were properly placed. Several retaining wall failures have been attributable to the dowels being on the wrong face of the wall!

Other inspections may be required by the building official, or by the EOR.

The Investigation

The geotechnical report for a project will nearly always have recommendations for site preparation (e.g. if fill is present or there is a liquefaction problem) in addition to design criteria information. This investigation report is usually a part of the contract documents and should be carefully reviewed and observed.

Forensic Investigations

If a problem is evident, or suspected, an independent engineer may be retained to investigate the problem. This will involve a review of the design, particularly to determine if the site conditions match the design criteria (e.g. a wall designed to retain eight feet, and actually retaining ten feet). The plans will be reviewed for clarity and conformance with the design intent and applicable building codes. The wall will be measured, deflection checked, and testing done to determine positioning of reinforcing and material strengths. Cores are often taken to determine both concrete strength and grout penetration into cells. A post-construction geotechnical report will be reviewed and perhaps more soil samples recommended.

When the cause of the problem is discovered, the most economical solution acceptable to the owner should be determined as part of the Forensic Investigation. This can be contentious, particularly if opposing parties offer different solutions. Hopefully the issues can be resolved equitably and with civility without resort to litigation. If an impasse, mediation can be a very effective and less costly resolution of a dispute.

24. RETAINING WALL DESIGN EXAMPLES

Description of Design Examples

These fourteen designs illustrate a variety of design conditions for retaining walls. They are worked by hand - the way you are accustomed to design retaining walls. You may use a different format, and your methodology may be a little different, but the results should be nearly the same. They are intended to show accepted design procedures. They are based on IBC '15, ACI 318-14, MSJC '05, and NMCR-SRW.

Following each of the examples is a report printout for the same problem using Retain Pro 10. This way you can compare results, which will closely agree, given round-offs and shortcuts in the hand calcs, which most of us do for expediency.

Example #1 – Retaining wall with sloped backfill, and stem of both concrete and masonry. The problem is designed so a key is necessary

Example #2 – A wall with an adjacent footing, and wind on a projecting stem

Example #3 – This problem illustrates a heel-side surcharge, and an axial load consisting of both dead and live load, and an eccentricity

Example #4 – This wall has a fence (zero weight and with wind load) on top of the retaining wall, and a property line condition

Example #5 – This is a freestanding wall with seismic force due to self-weight applied, and only minor earth retaining. It is set on a property line. Remember that for free-standing walls designed for seismic or wind, these loads can act in each direction, and if the controlling direction is not obvious, you may need to check the reversed too.

Example #6 – This illustrates a concrete stem with the inside face tapered (battered) and with a seismic force due to earth pressure

Example #7 – Masonry "basement" wall restrained laterally near the top

Example #8 – Concrete "basement" wall restrained laterally near the top

Example #9 – A rubble gravity wall design

Example #10 – A segmental wall (MSE) with geogrids

Example #11 – A segmental gravity wall -- no geogrids

Example #12 – A pier foundations option for Example #1

Example #13 – Soldier pile design – cantilevered

Example #14 – Gabion Wall (or multi-wythe large blocks

DESIGN EXAMPLE 1   **Page 1 of 11**

<u>Design Data</u>

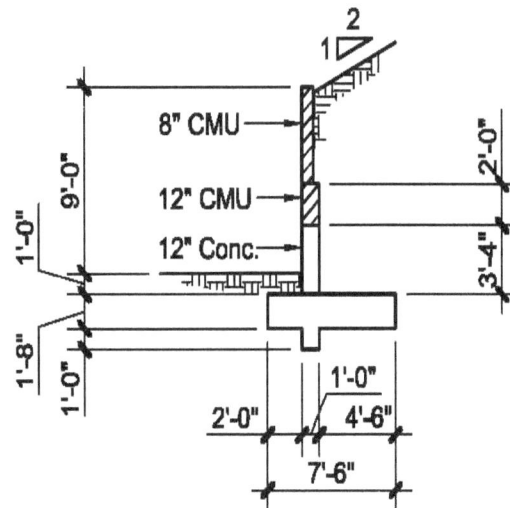

Code: IBC '15

Soil slope = 2:1

Soil density = 110 pcf

Use EPP (Equivalent fluid pressure method)

Equiv. fluid press. = 45 pcf

Active pressure toe side = 30 pcf

Passive press. = 389 pcf

$\mu = 0.40$

$F_s = 24{,}000$ psi

* $f'_m = 1500$ psi

$f'_c = 2500$ psi

$f_y = 60{,}000$ psi

Allow. soil pressure: 3000 psf

Alternatively check using Rankine

And assume angle of internal friction = 32°

for $\phi = 32°$ $\beta = \tan^{-1}\left(1/2\right) = 26.6° < 32°$

K_a (backfill slide horz. component) = 0.415   Pressure = .415 x 110 = 45.6 pcf

K_a (toe side assume $\phi = 34°$) = 0.28   Pressure = .28 x 110 = 30 pcf

Passive K_p (level $= \tan^2\left(45 + \dfrac{\varphi}{2}\right) = 3.54$ Pressure = 3.54 x 110 = 389 pcf

IBC '15 Load Factors for Strength Design (Concrete)

DL = 1.2
LL = 1.6
H   = 1.6
W   = 1.6
E   = 1.0

DESIGN EXAMPLE 1

Check Stem at Base

(Design Ht. = 0.00′)

$$M_u = \frac{45 \times 10^2}{2} \; x \; \frac{10}{3} \; x \; 1.6 \; = 11{,}999\,^{'\#}$$

Use b = 12″,   d = 9.6″

$$A_s \text{ (req'd)} = \frac{1.7\, f_c'\, bd}{2\, f_y} - \frac{1}{2} \sqrt{\frac{2.89\,(f_c'\, bd)^2}{f_y^2} - \frac{6.8\, f_c'\, bM_u}{\phi f_y^2}} = 0.29 \; sq.\ in.\ \text{(per CRSI}$$

eq.)

$$\rho_{bal.} = \frac{.85\, f_c'}{f_y} \; x \; .85\left(\frac{87{,}000}{87{,}000 + f_y}\right) = 0.018$$

$$\rho_{max.} = .75\, \rho_{bal.} = 0.0014 \; \rho_{min.} = \frac{200}{f_y} = .0033 \text{ (or at least 1.33 A}_s \text{ required per foot.)}$$

Try #6 at 16 $A_s = \dfrac{0.44}{1.33} = 0.33 > 0.29$ OK for strength

But $\rho = \dfrac{0.44}{16 \; x \; 9.6} = 0.0029 < .0033$ N.G.

Try #7 at 16″ $\rho = \dfrac{.60}{16 \; x \; 9.6} = .0039 > .0033$ As $= \dfrac{.60}{1.6/12} = .45$ sq. in. /ft.

OK

a   =   Asfy/.85 f_c' b = $\dfrac{.45 \times 60{,}000}{.85 \times 2500 \times 12} = 1.06$

$$\Phi M_n = .9 \times .45 \times 60{,}000 \left(9.6 - \frac{1.06}{2}\right) x \frac{1}{12} = 18{,}367\,^{'\#}$$

$$V_u \quad = \quad \left(\frac{45 \times 10^2}{2} - \frac{30 \times 1}{2}\right) 1.6 = 3576\#$$

$$v_u \quad = \quad \frac{3576}{12 \; x \; 9.6} = 31.0 \; v_{allow} = \phi \; 2\sqrt{f_c'} = 75 > 31.0 \text{ OK}$$

$$\underset{.75}{\uparrow}$$

DESIGN EXAMPLE 1   **Page 3 of 11**

Check embedment into footing:

* For hooked bar:

$$\frac{0.02 f_y\, d_b}{\sqrt{f_c'}} \; x \; 0.7 \; = \; \frac{.02 \; x \; 60{,}000 \; x \; 0.875 \; x \; 0.7}{\sqrt{2500}} \; = 14.7'' \quad or \; 8\,d_b \quad or \; 6''$$

* If ACI 318 '14, 25.4.10 is deemed applicable, reduce embedment by (A_s req'd) / (A_s provided).

  Min. footing thickness required = 14.7 + 3 = 17.7 in. Use 20″

Development length into stem

$$I_d \; = \; \frac{3\,F_y\,\alpha\beta\lambda\,d_b}{40\,\sqrt{f_c'}\left(\dfrac{c + K_{tr}}{d_b}\right)} \quad \textit{[See ACI 25.4.10}$$

$$\left(\frac{c + K_{tr}}{d_b}\right) = \left(\frac{2.0 + .44 + 0}{.875}\right) = 2.79 \; \rangle \; 2.5 \; max.$$

$$\therefore I_d = \frac{3 \; x \; 60{,}000 \; x \; 1 \; x \; 1 \; x \; 1 \; x \; .875}{40 \; x \; 50 \; x \; 2.5} = 31.5 \; in.$$

Note: If lapped with continuing reinf. of same size, splice length (assuming Class B splice) = 1.3 x 31.5 = 41″

<u>Check Masonry Stem</u> - Allowable Stress Design

Switch to 12″ masonry at 3′-4″ high which is approx. top of dowels.

$$f_m' = 1500 \; psi \qquad\qquad F_s - 24{,}000 \qquad\qquad f_b = .33 \; x \; 1500 = 500 \; psi$$

$$E_m = 900 \; f_m' \qquad E_m = 1{,}350{,}000 \; psi \qquad E_s = 29{,}000{,}000 \; n = \frac{E_s}{E_n} \; 21.5$$

$$M \; @ \; + 3.33' \; (= H \; of \; 6.67') = \frac{45 \; x \; 6.67^2}{2} \; x \frac{6.67}{3} = 2225^{'\#}$$

DESIGN EXAMPLE 1   **Page 4 of 11**

Try #5 at 16″ at edge

$$A_s = \frac{.31}{1.33} = 0.23 \qquad \rho = \frac{.31}{16 \times 9} = .0022$$

$$np = 21.5 \times .0022 = .0473 \qquad \frac{2}{kj} = 8.33 \quad j = 0.91 \quad \text{« From Amreim, Table E-9}$$

$$M_m = \frac{f_b \, bd^2}{\frac{2}{kj}} = \frac{500 \times 12 \times 9^2}{8.33} \times \frac{1}{12} = 4862^{'\#}$$

$$M_s = f_s \, A_s \, jd = 24,000 \times .23 \times .91 \times 9 \times \frac{1}{12} = 3767^{'\#} < 4862^{'\#} \text{ « governs}$$

∴   OK for # 5 @ 32 @ edge (overly conservative and could be reduced for final design)

$$V. = \frac{45 \times 6.67^2}{2} = 1001^{\#} \qquad v = \frac{1001}{12 \times .91 \times 9} = 10.2 < 1.0 \sqrt{f'_m} = 38.7 \quad \text{OK}$$

Lap length into concrete below (ACI '14, 25.4.2.3)

$$l_d = \frac{3 \times 60,000 \times 1 \times 1 \times 0.8 \times 0.625}{40 \times 50 \times 2.5} \times 1.3 = 23.4″$$

Lap length into masonry above = $.002 \, d_b f_s$

$= .002 \times .625 \times 24,000 = 30.0″$

Check Stem at + 5.33′

Reduce to 8″ masonry, grout reinf. cells only

$f'_m = 1500 \qquad F_s = 24,000 \quad F_b = 1500 \times .33 = 500 \text{ psi}$

$d = 5.25″ \qquad\qquad n = 21.5$

Depth @ + 5.33′ = 10.00 - 5.33 = 4.67′

$$M = \frac{45 \times 4.67^2}{2} \times \frac{4.67}{3} = 763^{'\#}$$

Use #5 at 32″ at edge

$$np = \frac{21.5 \times .31}{32 \times 5.25} = .04$$

DESIGN EXAMPLE 1 **Page 5 of 11**

$\dfrac{2}{kj} = 8.87$ $j = 0.92$

$$M_m = \frac{500 \times 12 \times 5.25^2}{8.87} \times \frac{1}{12} = 1553^{'\#}/ft$$

$$M_s = 24,000 \times \frac{.31}{2.67} \times .92 \times 5.25 \times \frac{1}{12} = 1121^{'\#}/ft \leftarrow governs$$

OK because 1121 > 763

$$V = \frac{45 \, x \, 6.67^2}{2} = 490^{\#}$$

$$v = \frac{490}{12 \times 0.92 \times 5.25} = 8.5 \, psi$$

$v_{allow} = 38.7 \, psi \, \rangle \, 8.5$ OK

Lap embedment below

$= .002 \, d_b \times 24,000$

$= .002 \times .625 \times 24,000 = 30''$

DESIGN EXAMPLE 1				Page 6 of 11

Stability Check

Item	Wt. (lbs.)	arm (ft.)	+M (ft.-lbs.)	-M (ft.-lbs.)
8″ CMU 78 x 4.67	364	2.33	849	
12″ CMU 124 x 2.0	248	2.50	620	
12″ conc. 150 x 3.33	500	2.50	1250	
earth 4.5 x 10 x 110	4,950	5.25	25,988	
earth .33 x 4.67 x 110	170	2.83	481	
earth 4.5 x 2.25 x 110 x ½	557	5.67	3,157	
earth 2.0 x 1.0 x 110	220	1.0	220	
footing 7.5 x 1.67 x 150	1,875	3.75	7,045	
key 1 x 1 x 150	150	2.5	375	
$P_a =$		$\frac{13.92}{3}$		20,229
$\frac{45 \times (10.0 + 1.67 + 2.25)^2}{2} = 4360^{\#/ft}$ – horizonta				
$P_v = ½ \times 4360$	2,180	7.5	16,350	
	11,214		56,335	20,229
w/o $P_v =$	9,034		39,985	

$$\bar{x} = \frac{RM - OTM}{W} = \frac{39,985 - 20,229}{9034} = 2.18'\quad \text{* About front edge of footing}$$

$$e = \frac{7.5}{2} - 2.18 = 1.56' = 18.7"$$

$$\text{Middle } \frac{1}{3} = \frac{7.5}{6} = 1.25 < 1.54 \qquad \therefore \text{ outside middle third}$$

When resultant <u>outside</u> middle third:

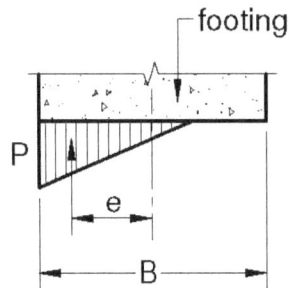

Soil pressure = $\dfrac{W}{.75B - 1.5e}$

Soil P = $\dfrac{9034}{.75 \ x \ 7.5 \ - \ 1.5 \ x \ 1.56}$ = 2725< 3000 lbs./sq. ft.   OK

OTM ratio = $\dfrac{39,985}{20,229}$ = 1.98

Check OTM Using P_v

RM   =   39,985 + 16,350 = 56,335 $^{'\#}$

OTM   20,229

OTM ratio = $\dfrac{56,635}{20,229}$ = 2.78

V   =   9034 + 2180   =   11,214$^{\#}$/ft

DESIGN EXAMPLE 1 **Page 8 of 11**

Check Sliding

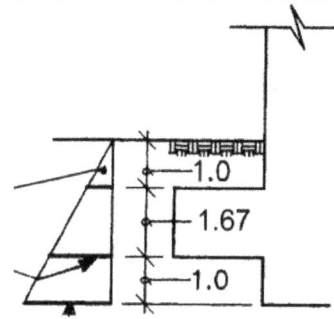

$$\text{Total lateral} = \frac{45 \times 13.92^2}{2} - \frac{30 \times 2.67^2}{2} = 4359^{\#}$$

Friction resistance = 9034 x 0.40 = 3614$^{\#}$

Passive resistance: Neglect

389 x 2.67 = 1039

389 x 3.67 = 1428

$$\text{Total passive} = 1428 \times 3.67 \times \tfrac{1}{2} - 389 \times 1 \times \tfrac{1}{2} = 2426^{\#}$$

$$\text{Factor of safety} = \frac{3614 + 2426}{4359} = 1.39 \approx 1.50$$

Consider OK

Check Key for Passive Resistance

$$V_u = \frac{1428 + 1039}{2} \times 1 \times 1.6 = 1974^{\#}$$

$$v_u = \frac{1974}{12(12-2)} = 16.4 \; < \; 2\varphi\sqrt{f_c'} = 65.1$$
$$\uparrow \qquad\qquad \uparrow = 0.65$$

(deduct 2″ from footing thickness for plain concrete)   [ACI '14, 14.5.1.7]

$$M_u = \left[1039 \times 1 \times \frac{12}{2} + \frac{(1428 - 1039)\,1.0}{2} \times \frac{2 \times 12}{3} \right] 1.6 = 12{,}464^{"\#}$$

$$S = \frac{12\,(12-2)^2}{6} = 200$$

$$f_r = \frac{12{,}464}{200} = 62.3\,\text{psi} \; < \; 5\varphi\sqrt{f_c'} = 5 \times 0.55 \times 50 = 137.5\,\text{psc} \quad \text{OK}$$

No reinf. req'd.

DESIGN EXAMPLE 1

Check Heel

Neglect upward soil pressure

$W_1 = 4950$

$W_2 = 557$ Use d = 20.0 - 2.5 = 17.5 in.

$W_3 = 4.5 \times 1.67 \times 150 = 1127^{\#}$

$W_4 = (P_v \text{ not used})$

$M_u = 4950 \times 2.25 \times 1.2 + 557 \times 3.0 \times 1.2 + 1127 \times 2.25 \times 1.2$

$= 18,413^{'\#}$

$Mu_{Stem} = 11,999^{'\#}$ ← governs

(Note: By statics, heel moment cannot exceed applied stem moment when upward soil pressure neglected.)

$$*A_s = \frac{1.7\, f_c'\, bd}{2\, f_y'} - \frac{1}{2}\sqrt{\frac{2.89\,(f_c'\, bd)^2}{f_y^2} - \frac{6.8\, f_b'\, M_u}{\varphi\, f_y^2}}$$

*[from C.R.S.I. Handbook, M_u in inch-kips]

ϕ = 0.90

$\downarrow$ ACI 9.6.1.3

A_s = 0.154 $sq.\,in./ft.$ or $\dfrac{200}{f_y} \times 12 \times 17.5 = .57$ or $1.33 \times 0.154 = 0.20$

Select #6 at 16 to match stem dowels (A_s = .33 > .20   OK)

Embedment length beyond stem face (using stress ratio reduction)

$$= \frac{3 \times 60,000 \times 1.0 \times 1.0 \times 0.8 \times 0.75}{40 \times 50 \times 2.5} \times .80 \times \frac{.154}{.44/1.33} = 8.04'' \langle 12''\min.$$

Use 12" embedment beyond stem face.

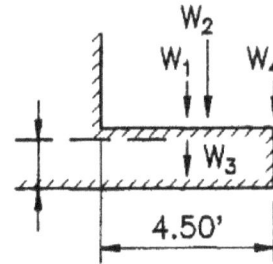

DESIGN EXAMPLE 1   **Page 10 of 11**

V_u = $4950 \times 1.2 + 557 \times 1.2 + 1127 \times 1.2 + (w/o \ P_v) = 7961\# / ft$

v_u = $\dfrac{7961}{12 \ x \ 17.5} = 37.9 \, psi < 2\varphi \sqrt{f_c'} = 2 \ x \ .75 \ x \sqrt{2500} = 75 \, psi$ OK

Check Toe

Apply total <u>factored</u> vertical load

(w/o p_v) at same eccentricity as

service loads:

     = $9034 \times 1.2 = 10,841^{\#}$

Factored soil pressure

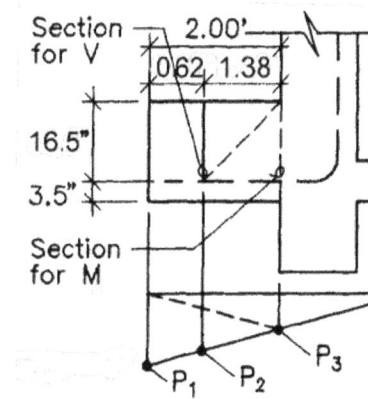

     = $\dfrac{9034}{.75 \ x \ 7.5 - 1.5 \ x \ 1.56} = 2725 \, psf$ @ P_1

Soil Pressure @ P_2 (for shear)

     = $\dfrac{3 \ x \ 2.21 - .62}{3 \ x \ 2.21} \ x \ 3270 = 2964 \ psf$

Soil Pressure @ P_3 (Max. M)   =

     = $\dfrac{3 \ x \ 2.21 - 2.0}{3 \ x \ 2.21} \ x \ 3270 = 2283 \ psf$

Because stem moment governs:

V_u = $\dfrac{11,992}{4.5/2} \ x \ \dfrac{1.2}{1.6} = 3997^{\#}$

v_u = $3997 / (12 \times 17.5) = 19.0 \ psi \ < 75 \ \ OK$

DESIGN EXAMPLE 1   **Page 11 of 11**

$$M_u\uparrow \;\; = \;\; \frac{3270\,x\,2}{2}\,x\,.67\,x\,2 + \frac{2284\,x\,2}{2}\,x\,.33\,x\,2 = 5889^{'\#}$$

$$M_u\downarrow \;\; = \;\; (1.67\,x\,150 + 1\,x\,110)\,x\,\frac{2^2}{2}\,x\,1.2 = 865^{'\#}$$

$$M_{design} \;\; = \;\; 5889 - 865 = 5024^{'\#}$$

$$A_s min = \frac{200}{f_y}\,x\,12\,x\,16.5 = .66 \text{ sq. in. / ft}$$

A_s required (per C.R.S.I equation) = .07 sq. in.

A_s to override A_s min = 1.33x .07 = 0.09 sq. in.

But also As min = 0.0018 x 12 x 16.5 = 0.36 $\leftarrow$ Governs

Use #7 @ 16" ($A_s = \dfrac{0.60}{1.33} = 0.45 > 0.36$)   OK

Shear negligible by inspection because only acts on 0.62 ft.

DESIGN EXAMPLE 1   Report Printout

Basics of Retaining Wall Design   Title   EX-1 :   Page : 1

Job # :   Dsgnr: HB   Date: 9 FEB 2018

Description....

EX-1

This Wall in File: C:\Users\chris\Dropbox\RetainPro 10 Project Files\Hugh.RPX

RetainPro (c) 1987-2018, Build 11.16.1.16
License : KW-06050001
License To : RETAIN PRO SOFTWARE

Cantilevered Retaining Wall   Code: IBC 2015,ACI 318-14,ACI 530-13

Criteria

Retained Height	=	10.00 ft
Wall height above soil	=	0.00 ft
Slope Behind Wall	=	2.00
Height of Soil over Toe	=	12.00 in
Water height over heel	=	0.0 ft

Soil Data

Allow Soil Bearing	=	3,000.0 psf
Equivalent Fluid Pressure Method		
Active Heel Pressure	=	45.0 psf/ft
Passive Pressure	=	389.0 psf/ft
Soil Density, Heel	=	110.00 pcf
Soil Density, Toe	=	110.00 pcf
Footing||Soil Friction	=	0.400
Soil height to ignore for passive pressure	=	12.00 in

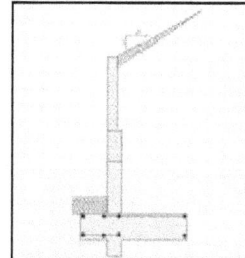

Surcharge Loads

Surcharge Over Heel	=	0.0 psf
Used To Resist Sliding & Overturning		
Surcharge Over Toe	=	0.0 psf
Used for Sliding & Overturning		

Axial Load Applied to Stem

Axial Dead Load	=	0.0 lbs
Axial Live Load	=	0.0 lbs
Axial Load Eccentricity	=	0.0 in

Lateral Load Applied to Stem

Lateral Load	=	0.0 #/ft
...Height to Top	=	0.00 ft
...Height to Bottom	=	1.00 ft
Load Type	=	Wind (W)
		(Service Level)
Wind on Exposed Stem (Service Level)		0.0 psf

Adjacent Footing Load

Adjacent Footing Load	=	0.0 lbs
Footing Width	=	0.00 ft
Eccentricity	=	0.00 in
Wall to Ftg CL Dist	=	0.00 ft
Footing Type		Line Load
Base Above/Below Soil at Back of Wall	=	0.0 ft
Poisson's Ratio	=	0.300

Design Summary

Wall Stability Ratios

Overturning	=	1.99 OK
Sliding	=	1.38 Ratio < 1.5!
Total Bearing Load	=	9,036 lbs
...resultant ecc.	=	18.51 in
Soil Pressure @ Toe	=	2,729 psf OK
Soil Pressure @ Heel	=	0 psf OK
Allowable	=	3,000 psf
Soil Pressure Less Than Allowable		
ACI Factored @ Toe	=	3,820 psf
ACI Factored @ Heel	=	0 psf
Footing Shear @ Toe	=	9.2 psi OK
Footing Shear @ Heel	=	37.9 psi OK
Allowable	=	75.0 psi

Sliding Calcs

Lateral Sliding Force	=	4,357.7 lbs
less 100% Passive Force	=	- 2,420.4 lbs
less 100% Friction Force	=	- 3,613.9 lbs
Added Force Req'd	=	0.0 lbs OK
...for 1.5 Stability	=	502.1 lbs NG

Vertical component of active lateral soil pressure IS
NOT considered in the calculation of soil bearing

Load Factors

Building Code	IBC 2015,ACI
Dead Load	1.200
Live Load	1.600
Earth, H	1.600
Wind, W	1.000
Seismic, E	1.000

Stem Construction

			3rd	2nd	Bottom
			Stem OK	Stem OK	Stem OK
Design Height Above Ftg	ft =		5.33	3.33	0.00
Wall Material Above "Ht"	=		Masonry	Masonry	Concrete
Design Method	=		ASD	ASD	LRFD
Thickness	=		8.00	12.00	12.00
Rebar Size	=		# 5	# 5	# 7
Rebar Spacing	=		32.00	16.00	16.00
Rebar Placed at	=		Edge	Edge	Edge

Design Data

			3rd	2nd	Bottom
fb/FB + fa/Fa	=		0.611	0.437	0.656
Total Force @ Section					
Service Level	lbs =		490.7	1,001.0	
Strength Level	lbs =				3,600.0
Moment....Actual					
Service Level	ft-# =		763.9	2,225.6	
Strength Level	ft-# =				12,000.0
Moment.....Allowable	ft-# =		1,494.8	5,093.8	18,288.8
Shear.....Actual					
Service Level	psi =		5.4	7.2	
Strength Level	psi =				31.4
Shear.....Allowable	psi =		44.8	45.0	75.0
Anet (Masonry)	in2 =		91.50	139.50	
Rebar Depth 'd'	in =		5.25	9.00	9.56

Masonry Data

			3rd	2nd	Bottom
fm	psi =		1,500	1,500	
Fs	psi =		32,000	32,000	
Solid Grouting	=		Yes	Yes	
Modular Ratio 'n'	=		21.48	21.48	
Wall Weight	psf =		78.0	124.0	150.0
Short Term Factor	=		1.000	1.000	
Equiv. Solid Thick.	in =		7.60	11.60	
Masonry Block Type	=		Medium Weight		
Masonry Design Method	=		ASD		

Concrete Data

			3rd	2nd	Bottom
f'c	psi =				2,500.0
Fy	psi =				60,000.0

DESIGN EXAMPLE 1   Report Printout

Basics of Retaining Wall Design	Title: EX-1		Page: 2
	Job #:	Dsgnr: HB	Date: 2 JUL 2018
	Description:		
	EX-1		

This Wall in File: c:\Users\chris\Dropbox\RetainPro 10 Project Files\HB\Hugh.RPX

RetainPro (c) 1987-2018, Build 11.18.08.22
License : KW-06000216
License To : RetainPro Development Center

Cantilevered Retaining Wall   Code: IBC 2015, ACI 318-14, ACI 530-13

Concrete Stem Rebar Area Details

Bottom Stem	Vertical Reinforcing	Horizontal Reinforcing	
As (based on applied moment) :	0.2885 in2/ft		
(4/3) * As	0.3846 in2/ft	Min Stem T&S Reinf Area 0.959 in2	
200bd/fy : 200(12)(9.5625)/60000	0.3825 in2/ft	Min Stem T&S Reinf Area per ft of stem Height 0.288 in2/ft	
0.0018bh : 0.0018(12)(12)	0.2592 in2/ft	Horizontal Reinforcing Options :	
	------------	One layer of	Two layers of
Required Area	0.3825 in2/ft	#4 @ 9.33 in	#4 @ 16.67 in
Provided Area	0.45 in2/ft	#5 @ 12.92 in	#5 @ 25.83 in
Maximum Area	1.2964 in2/ft	#6 @ 18.33 in	#6 @ 36.67 in

Footing Dimensions & Strengths

Toe Width	=	2.00 ft
Heel Width	=	5.50
Total Footing Width	=	7.50
Footing Thickness	=	20.00 in
Key Width	=	12.00 in
Key Depth	=	12.00 in
Key Distance from Toe	=	2.00 ft

f'c = 2,500 psi   Fy = 60,000 psi
Footing Concrete Density = 150.00 pcf
Min. As % 0.0018
Cover @ Top 2.00 @ Btm = 3.00 in

Footing Design Results

		Toe	Heel
Factored Pressure	=	4,060	0 psf
Mu': Upward	=	7,293	0 ft-#
Mu': Downward	=	864	21,407 ft-#
Mu: Design	=	6,429	21,407 ft-#
Actual 1-Way Shear	=	9.81	40.96 psi
Allow 1-Way Shear	=	40.00	75.00 psi
Toe Reinforcing	=	#7 @ 16.00 in	
Heel Reinforcing	=	#6 @ 16.00 in	
Key Reinforcing	=	None Spec'd	

Other Acceptable Sizes & Spacings
Toe: Not req'd: Mu < phi*5*lambda*sqrt(fc)*Sm
Heel: #4 @ 5.56 in, #5 @ 8.61 in, #6 @ 12.22 in, #7 @ 16.67 in, #8 @ 21.94 in, #9 @ 27
Key: Not req'd: Mu < phi*5*lambda*sqrt(fc)*Sm

Min footing T&S reinf Area	3.24 in2
Min footing T&S reinf Area per foot	0.43 in2/ft

If one layer of horizontal bars: If two layers of horizontal bars:
#4 @ 5.56 in   #4 @ 11.11 in
#5 @ 8.61 in   #5 @ 17.22 in
#6 @ 12.22 in   #6 @ 24.44 in

Summary of Overturning & Resisting Forces & Moments

Item	OVERTURNING Force lbs	Distance ft	Moment ft-#		RESISTING Force lbs	Distance ft	Moment ft-#
Heel Active Pressure =	4,462.7	4.69	20,949.7	Soil Over Heel =	5,316.7	5.08	27,026.4
Surcharge over Heel =				Sloped Soil Over Heel =	642.4	5.89	3,783.2
Surcharge Over Toe =				Surcharge Over Heel =			
Adjacent Footing Load =				Adjacent Footing Load =			
Added Lateral Load =				Axial Dead Load on Stem =			
Load @ Stem Above Soil =				* Axial Live Load on Stem =			
				Soil Over Toe =	220.0	1.00	220.0
				Surcharge Over Toe =			
				Stem Weight(s) =	1,111.8	2.45	2,718.7
Total	4,462.7	O.T.M.	20,949.7	Earth @ Stem Transitions =	171.2	2.83	485.2
				Footing Weight =	1,875.0	3.75	7,031.3
Resisting/Overturning Ratio	=	1.99		Key Weight =	150.0	2.50	375.0
Vertical Loads used for Soil Pressure =	9,487.1 lbs			Vert. Component =			
				Total =	9,487.1 lbs	R.M. =	41,639.7

* Axial live load NOT included in total displayed, or used for overturning
resistance, but is included for soil pressure calculation.

Vertical component of active lateral soil pressure IS NOT considered in
the calculation of Sliding Resistance.

Vertical component of active lateral soil pressure IS NOT considered in
the calculation of Overturning Resistance.

DESIGN EXAMPLE 1   Report Printout

Basics of Retaining Wall Design

Title   EX-1
Job #
Description...
EX-1

Dsgnr   HB

Date   2 JUL 2018

Page : 3

This Wall in File: c:\Users\chris\Dropbox\RetainPro 10 Project Files\HB\Hugh.RPX

RetainPro (c) 1987-2018, Build 11.18.08.22
License : KW-06000210
License To : RetainPro Development Center

Cantilevered Retaining Wall

Code: IBC 2015,ACI 318-14,ACI 530-13

| Tilt |

Horizontal Deflection at Top of Wall due to settlement of soil

(Deflection due to wall bending not considered)

Soil Spring Reaction Modulus	250.0	pci
Horizontal Defl @ Top of Wall (approximate only)	0.107	in

The above calculation is not valid if the heel soil bearing pressure exceeds that of the toe, because the wall would then tend to rotate into the retained soil.

DESIGN EXAMPLE 2   **Page 1 of 8**

<u>Design Data</u>

    Building Code: IBC '15

    Soil bearing = 4000 psf

    Soil density = 110 pcf

    Eq. fluid pressure = 30 pcf

    Passive = 300 pcf

    μ = 0.40

    Poisson's Ratio (for Bousinesq) = 0.50

    f'_m = 1,500 psi

    f_s = 24,000 psi

    f_y = 60,000 psi

    f'_c = 2,000 psi

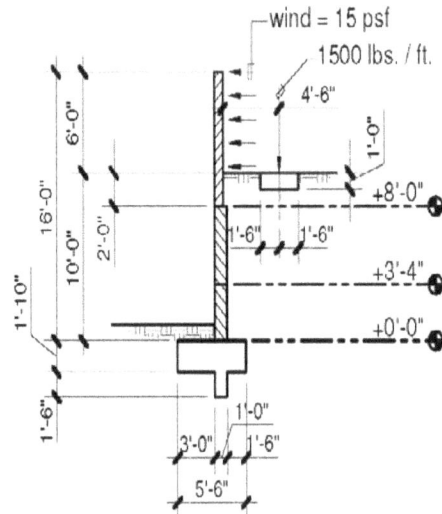

<u>Check Stem at +8.00'</u>

$$M = 15 \times 6 \left(\frac{6}{2} + 2 \right) + \frac{30 \times 2^2}{2} \, x \, \frac{2}{3}$$

$$= 490^{'\#}$$

f'_m = 1500 psi $\qquad$ f_b = .33 x 1500 = 500 psi

$\qquad$ d = 3.75 (for 8" CMU)

$E_m = 900 \times f'_m = 1,350,000$ psi

$E_s = 29,000,000$ psi $\qquad$ n = 21.5

$$A_s = \frac{490 \times 12}{24,000 \times 1.33 \times .90 \times 3.75} = .054 \quad \underline{\text{Use \#4 @ 32}}$$

$$\text{Assume} \uparrow \qquad A_s = \frac{.20}{2.67} = .075 \quad \text{OK}$$

$$np = \frac{21.5 \times .20}{32 \times 3.75} = .036 \qquad \frac{2}{kj} = 9.24 \qquad j = .92$$

DESIGN EXAMPLE 2 **Page 2 of 8**

$$M_s = 24{,}000 \times 3 \times .075 \times .92 \times 3.75 \times \frac{1}{12} = 518.8^{'\#} \leftarrow \text{ governs}$$

$$M_M = 500 \times 12 \times 3.75^2 \times \frac{1}{9.24} \times \frac{1}{12} = 761^{'\#}$$

$$V = 6' \times 15 + \frac{30 \times 2^2}{2} = 150^{\#}$$

$$\nu = \frac{150}{12 \times .92 \times 3.75} = 3.62 \text{ psi } \nu_{allow} = \sqrt{f'_c} \times 1.33 = 59 \text{ psi}$$

Rebar embedment below = .002 x .50 x 24,000 = 24"

Check Stem @ + 3.33'

Change to 12" CMU, d = 9.0"

$$M_{wind} = 6' \times 15 \text{ psf} \times \left(\frac{6}{2} + 6.67 \right) = 870^{'\#}$$

$$M_{soil} = \frac{30 \times 6.67^2}{2} \times \frac{6.67}{3} = 1484^{'\#}$$

$$M_{bousinesq} \text{ (from program)} = \underline{1107}$$
$$3461^{'\#}$$

$$\text{Total lateral} = 6' \times 15 + \frac{30 \times 6.67^2}{2} + 505 = 1262^{\#} / \text{ft of wall}$$
$$\text{(from program)} \uparrow$$

$$A_s = \frac{3461 \times 12}{24{,}000 \times .9 \times 9.0} = 0.21 / \text{sq in} \qquad \text{Use \#5 @ 16 @ edge}$$

$$\left(\text{Disallow } \frac{1}{3} \text{ wind stress} \right.$$
$$\left. \text{increase at this level} \right)$$

$$np = \frac{21.5 \times .31}{16 \times 9} = .046 \qquad \frac{2}{kj} = 8.4 \qquad j = 0.91 \qquad A_s = \frac{.31}{1.33} = .23 / \text{sq in}$$

$$M_s = .23 \times 24{,}000 \times .91 \times 9 \times \frac{1}{12} = 3767^{'\#} \qquad > 3461 \text{ OK}$$

DESIGN EXAMPLE 2	Page 3 of 8

$$M_m = \frac{500 \times 12 \times 9^2}{8.4} \times \frac{1}{12} = 4821 \, '^{\#}$$

$$V = 6 \times 15 + \frac{30 \times 6.67^2}{2} + 504 = 1261\#$$

$$v = \frac{1261}{12 \times .91 \times 9} = 12.8 \; < \sqrt{f'_m} = 38.7 \quad OK$$

Embedment length = .002 x .625 x 24,000 = 30"

Check Stem @ Base

Use 12" CMU

$$M_{wind} = 6' \times 15 \, psf \left(\frac{6}{2} + 10\right) = 1170 \, '^{\#}$$

$$M_{soil} = \frac{30 \times 10^2}{2} \times \frac{10}{3} = 5000$$

$$M_{bousiresq.} \text{ (from program)} = \underline{3245}$$
$$9415 \, '^{\#}$$

$$f'_m = 1500 \, psi.$$

$$f_b = 500 \, psi \quad n = 21.5 \quad d = 9.0 \quad A_s = \frac{9415 \times 12}{24,000 \times .90 \times 9} = 0.58$$

↑ Assume

Try #8 @ 8"

$$A_s = \frac{0.79}{0.67} = 1.18 \text{ sq. in./ft.}$$

$$np = \frac{21.5 \times 0.79}{8 \times 9} = 0.236 \qquad \frac{2}{kj} = 4.88 \qquad j = .83$$

DESIGN EXAMPLE 2

$$M_s = \frac{1.33 \; x \; 24,000 \; x \; .83 \; x \; 9.0}{12} = 19,870"\#$$

$$M_m = \frac{500 \; x \; 12 \; x \; 9^2}{4.88} \; x \; \frac{1}{12} = 8299$$

Stress Ratio $= \dfrac{9415}{8299} = 1.13$

(13% overstressed –optional redesign)

$$V = 6' \; x \; 15 + \frac{30 \; x \; 10^2}{2} + 737 = 2327^{\#} / \text{ft}$$

$$\text{Bousinesq} \uparrow$$

$$v = \frac{2327}{12 \; x \; 9} = 21.6 < \sqrt{f'_m} = 38.7 \quad \text{OK}$$

Development length of dowels

ℓ_{db} = .002 x 24,000 x 1.0 x 1.5 = 72.0 in.

Choose not to reduce by stress ratio. Assume continuing bars will be smaller diameter, therefore, 1.3 multiplier for splice lap not required.

Embedment for hooked bar into footing $= \dfrac{0.02 \; x \; f_y \, d_b \; x \; 0.7}{\sqrt{f'_c}}$

$$= \frac{.02 \; x \; 60,000 \; x \; 1.0 \; x \; 0.7}{44.7} = 18.8"$$

Min. ftg. Thickness = 18.8 + 3.0 = 21.5"

Use 22" thick

DESIGN EXAMPLE 2   **Page 5 of 8**

Stability

Try footing 5'-6" x 1'-10" thick

Item	Wt.		arm		+M	-M
8" stem 8' x 78 psf	624	x	3.33	=	2078	
12" stem 8' x 124 psf	992	x	3.50	=	3472	
Soil @ heel 1.5 x 110 x 10'	1650	x	4.75	=	7838	
Soil @ toe 3 x 1 x 110	330	x	1.50	=	495	
Soil behind stem .33 x 110 x 2	74	x	3.83	=	283	
Footing 5.5 x 1.83 x 150	1510	x	2.75	=	4152	
Key	225	x	3.50	=	788	
Adj. Footing	167	x	4.75	=	795	
OTM Wind = 6' x 15			14.83			1335

$$\text{OTM soil} = \frac{30 \; x \; 11.83^2}{2} \qquad \frac{11.83}{3} \qquad\qquad 8278$$

OTM adj. footing 830#			5.7			4731
	5572#		$\frac{2.83}{3}$		19,899$^{\#}$	14,344$^{\#}$

$$\bar{x} = \frac{19{,}889 - 14{,}344}{5572} = 1.00$$

$$e = \frac{5.5}{2} - 1.00 = 1.75 > \frac{5.5}{6} = .92$$

$\therefore$ outside middle third

$$\text{Soil p} = \frac{5572}{.75 \; x \; 5.5 - 1.5 \; x \, 1.75} = 3715 \; psf \; \langle \; 4000 \qquad \text{OK}$$

$$\text{OTM ratio} = \frac{19{,}899}{14{,}344} = 1.39 \; \approx \; 1.50 \quad \text{(consider OK?)}$$

DESIGN EXAMPLE 2

Check Sliding

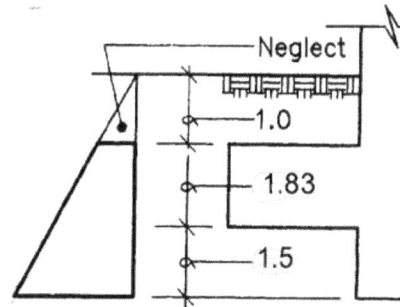

Lateral force of adj footing

$$\text{Total lateral} = 15 \times 6' + \frac{30 \times 11.83^2}{2} + 830$$

$$= 3019 \text{ lbs.}$$

Friction resistance

$$= 5572 \times .40 = 2229 \text{ lbs}$$

Passive resistance

$$=$$

$$\frac{300 \times 4.33^2}{2} - \frac{300 \times 1^2}{2} = 2662 \text{ lbs}$$

$$\text{Sliding ratio} = \frac{2229 + 2662}{3019} = 1.62$$

Check Key

$$\text{Force} = \frac{300 \times 4.33^2}{2} - \frac{300 \times 2.83^2}{2} = 1611^{\#}$$

Effective width = 12" - 2" = 10"

$$v = \frac{1.6 \times 1611}{12 \times 10} = 23.9 < 2\varphi \sqrt{f_c'} = 85 \quad \text{OK}$$

$$M_u \text{ (approx.)} = 1611 \times \frac{18''}{2} \times 1.6 = 23,198^{"\#}$$

DESIGN EXAMPLE 2   **Page 7 of 8**

$$S = \frac{12 \ x \ (12-2)^2}{6} = 200$$

$$f_t = \frac{23,198}{200} = 116 < 5\varphi \sqrt{f_c'} = 137$$

OK

$\therefore$ No reinforcing required

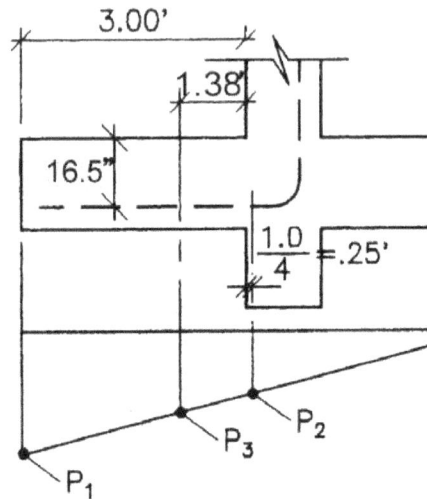

<u>Check Toe</u>

Total vert. factored load
$$= 5572 \ x \ 1.2 = 6686^{\#}$$

Factored soil pressure

$$P_1 = \frac{6686}{.75 \ x \ 5.5 - 1.5 \ x \ 1.73} = 4370 \ psf \ @$$

$$P_2 = \frac{3 \ x \ 1.02 - 3.0 - .25}{3 \ x \ 1.02} \ x \ 4370 \cong 0.0$$

$$P_3 = \frac{3 \ x \ 1.02 - 3.0 + 1.54}{3 \ x \ 1.02} \ x \ 4370 = 2285 \ psf$$

$$M_u\uparrow = 4370 \ x \ 3.25 \ x \ \frac{1}{2} \ x \ .67 \ x \ 3.25 = 15,394 \ '^{\#}$$

$$M_u\downarrow = \left[(1.83 \ x \ 150 + 1 \ x \ 110) \ x \ 3 \ x \left(\frac{3}{2} + .25\right) \right] x \ 1.2 = 2,422 \ '^{\#}$$

$$M_{design} = 15,394 - 2,422 = 12,972 \ '^{\#}$$

A_s (Per CRSI equation – See Example #1) = 0.17

$$or \left(\frac{200}{f_y}\right) 10 \ x \ 10.5 = 0.35 \ sq. \ in./ft.$$

Use 1.33 x 0.17 = 0.23

Use dowel bars = #8 @ 8" $A_s = \frac{.79}{.67} = 1.18$ OK

$$V_u = \frac{4370 + 2285}{2} \ x \ (3 - 1.54) - (1.83 \ x \ 150 + 1 \ x \ 110)(3 - .1.54) \ 1.2$$

$$= 4996\#$$

DESIGN EXAMPLE 2   **Page 8 of 8**

$$v_u = \frac{4996}{12 \ x \ 18.5} = 22.5$$

$$v_{allow} = 2 \ x \ .85 \ \sqrt{2500} = 85 \ OK$$

<u>Check Heel</u>

Ignore upward soil pressure

$$
\begin{aligned}
W_1 &= 992 \ x \ 1.2 = & 1190^{\#} \\
&+ 624 \ x \ 1.2 = & 749 \\
&+ 74 \ x \ 1.2 = & 89 \\
&+ 167 \ x \ 1.2 = & 200 \\
W^2 &= 1.83 \ x \ 1.5 \ x \ 150 \ x \ 1.2 = & \underline{494} \\
& & 2722^{\#}
\end{aligned}
$$

$$M_u = 3163 \ x \ \frac{1.5}{2} = 2372'^{\#}$$

Effective thickness of footing = 20 - 2 = 18"

$$S = \frac{12 \ x \ 18^2}{6} = 648$$

$$f_r = \frac{2372 \ x \ 12}{648} = 43.9 < 5\phi \sqrt{f_c'} = 137 \quad OK$$

∴   No reinforcement required

$$V_u = 2722 \ \#$$

$$v_u = \frac{2722}{12 \ x \ 20} = 11.3 < 2\phi \sqrt{f_c'} = 67 \quad OK$$

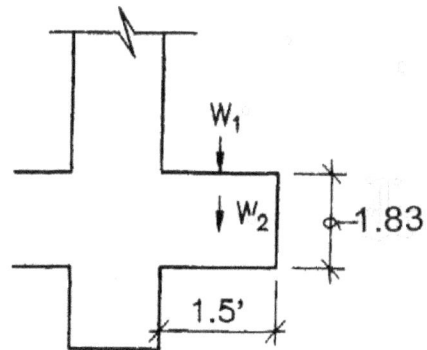

DESIGN EXAMPLE 2 | Report Printout

Basics of Retaining Wall Design
Title EX-2
Job #
Description
EX-2
Dsgnr HB
Date 2 JUL 2018
Page 1

This Wall in File: c:\Users\chris\Dropbox\RetainPro 10 Project Files\HB\Hugh.RPX
RetainPro (c) 1987-2018. Build 11.18.08.22
License : KW-06000216
License To : RetainPro Development Center

Cantilevered Retaining Wall

Code: IBC 2015, ACI 318-14, ACI 530-13

Criteria

Retained Height	=	10.00 ft
Wall height above soil	=	6.00 ft
Slope Behind Wall	=	0.00
Height of Soil over Toe	=	12.00 in
Water height over heel	=	0.0 ft

Soil Data

Allow Soil Bearing	=	4,000.0 psf
Equivalent Fluid Pressure Method		
Active Heel Pressure	=	30.0 psf/ft
Passive Pressure	=	300.0 psf/ft
Soil Density, Heel	=	110.00 pcf
Soil Density, Toe	=	110.00 pcf
Footing||Soil Friction	=	0.400
Soil height to ignore for passive pressure	=	12.00 in

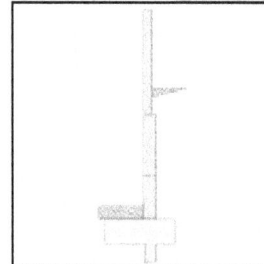

Surcharge Loads

Surcharge Over Heel	=	0.0 psf
Used To Resist Sliding & Overturning		
Surcharge Over Toe	=	0.0 psf
Used for Sliding & Overturning		

Axial Load Applied to Stem

Axial Dead Load	=	0.0 lbs
Axial Live Load	=	0.0 lbs
Axial Load Eccentricity	=	0.0 in

Lateral Load Applied to Stem

Lateral Load	=	0.0 #/ft
Height to Top	=	0.00 ft
Height to Bottom	=	0.00 ft
Load Type	=	Wind (W) (Service Level)
Wind on Exposed Stem (Service Level)	=	15.0 psf

Adjacent Footing Load

Adjacent Footing Load	=	1,500.0 lbs
Footing Width	=	3.00 ft
Eccentricity	=	0.00 in
Wall to Ftg CL Dist	=	4.50 ft
Footing Type	=	Line Load
Base Above/Below Soil at Back of Wall	=	-1.0 ft
Poisson's Ratio	=	0.500

Design Summary

Wall Stability Ratios

Overturning	=	1.49 Ratio < 1.5!
Sliding	=	1.60 OK
Total Bearing Load	=	5,952 lbs
resultant ecc.	=	18.80 in
Soil Pressure @ Toe	=	3,353 psf OK
Soil Pressure @ Heel	=	0 psf OK
Allowable	=	4,000 psf
Soil Pressure Less Than Allowable		
ACI Factored @ Toe	=	4,694 psf
ACI Factored @ Heel	=	0 psf
Footing Shear @ Toe	=	29.8 psi OK
Footing Shear @ Heel	=	12.6 psi OK
Allowable	=	67.1 psi

Sliding Calcs

Lateral Sliding Force	=	3,021.1 lbs
less 100% Passive Force	=	2,454.2 lbs
less 100% Friction Force	=	2,380.9 lbs
Added Force Req'd	=	0.0 lbs OK
for 1.5 Stability	=	0.0 lbs OK

Vertical component of active lateral soil pressure IS NOT considered in the calculation of soil bearing

Load Factors

Building Code	IBC 2015 ACI
Dead Load	1.200
Live Load	1.600
Earth, H	1.600
Wind, W	1.000
Seismic, E	1.000

Stem Construction

			3rd	2nd	Bottom
			Stem OK	Stem OK	Stem OK
Design Height Above Ftg	ft	=	8.00	3.33	0.00
Wall Material Above "Ht"		=	Masonry	Masonry	Masonry
Design Method		=	ASD	ASD	ASD
Thickness		=	8.00	12.00	12.00
Rebar Size		=	# 4	# 5	# 8
Rebar Spacing		=	32.00	16.00	8.00
Rebar Placed at		=	Center	Edge	Edge

Design Data

		3rd	2nd	Bottom
fb/FB + fa/Fa	=	0.713	0.600	0.940
Total Force @ Section				
Service Level	lbs =	162.1	1,261.5	2,341.9
Strength Level	lbs =			3,600.0
Moment...Actual				
Service Level	ft-# =	493.1	3,461.5	9,415.0
Strength Level	ft-# =			12,000.0
Moment...Allowable	ft-# =	691.5	5,093.8	11,211.4
Shear....Actual				
Service Level	psi =	1.8	9.0	16.8
Strength Level	psi =			31.4
Shear...Allowable	psi =	45.6	46.5	47.4
Anet (Masonry)	in2 =	91.50	139.50	139.50
Rebar Depth 'd'	in =	3.75	9.00	9.00

Masonry Data

		3rd	2nd	Bottom
f'm	psi =	1,500	1,500	1,500
Fs	psi =	32,000	32,000	32,000
Solid Grouting	=	Yes	Yes	Yes
Modular Ratio 'n'	=	21.48	21.48	21.48
Wall Weight	psf =	78.0	124.0	124.0
Short Term Factor	=	1.000	1.000	1.000
Equiv. Solid Thick	in =	7.60	11.60	11.60
Masonry Block Type	=	Medium Weight		
Masonry Design Method	=	ASD		

Concrete Data

Fc	psi =	
Fy	psi =	

DESIGN EXAMPLE 2   Report Printout

Basics of Retaining Wall Design

Title EX-2
Job #
Description
EX-2

Dsgnr: HB   Date: 2 JUL 2018   Page 2

This Wall in File: c:\Users\chris\Dropbox\RetainPro 10 Project Files\HB\Hugh.RPX

RetainPro (c) 1987-2018, Build 11.18.08.22
License : KW-06000216
License To : RetainPro Development Center

Cantilevered Retaining Wall   Code: IBC 2015, ACI 318-14, ACI 530-13

Footing Dimensions & Strengths

Toe Width	=	3.00 ft
Heel Width	=	2.50
Total Footing Width	=	5.50
Footing Thickness	=	22.00 in
Key Width	=	12.00 in
Key Depth	=	18.00 in
Key Distance from Toe	=	3.00 ft
f'c = 2,000 psi Fy =		60,000 psi
Footing Concrete Density =		150.00 pcf
Min. As %	=	0.0018
Cover @ Top 2.00 @ Btm =		3.00 in

Footing Design Results

		Toe	Heel
Factored Pressure	=	4,694	0 psf
Mu' : Upward	=	15,174	0 ft-#
Mu' : Downward	=	2,426	3,529 ft-#
Mu: Design	=	12,749	3,529 ft-#
Actual 1-Way Shear	=	29.80	12.60 psi
Allow 1-Way Shear	=	67.08	35.78 psi
Toe Reinforcing	=	None Spec'd	
Heel Reinforcing	=	None Spec'd	
Key Reinforcing	=	None Spec'd	

Other Acceptable Sizes & Spacings

Toe: #4@ 5.05 in, #5@ 7.83 in, #6@ 11.11 in, #7@ 15.15 in, #8@ 19.95 in, #9@ 25.
Heel: Not req'd: Mu < phi*5*lambda*sqrt(f'c)*Sm
Key: Not req'd: Mu < phi*5*lambda*sqrt(f'c)*Sm

Min footing T&S reinf Area   2.61 in2
Min footing T&S reinf Area per foot   0.48 in2 ft

If one layer of horizontal bars	If two layers of horizontal bars
#4@ 5.05 in	#4@ 10.10 in
#5@ 7.83 in	#5@ 15.66 in
#6@ 11.11 in	#6@ 22.22 in

Summary of Overturning & Resisting Forces & Moments

Item		OVERTURNING Force lbs	Distance ft	Moment ft-#			RESISTING Force lbs	Distance ft	Moment ft-#
Heel Active Pressure	=	2,100.4	3.94	8,285.0	Soil Over Heel	=	2,016.7	4.58	9,243.1
Surcharge over Heel	=				Sloped Soil Over Heel	=			
Surcharge Over Toe	=				Surcharge Over Heel	=			
Adjacent Footing Load	=	830.7	5.66	4,700.1	Adjacent Footing Load	=	203.7	4.58	933.6
Added Lateral Load	=				Axial Dead Load on Stem	=			
Load @ Stem Above Soil	=	90.0	14.83	1,335.0	* Axial Live Load on Stem	=			
					Soil Over Toe	=	330.0	1.50	495.0
					Surcharge Over Toe	=			
Total		3,021.1	O.T.M.	14,320.1	Stem Weight(s)	=	1,616.0	3.44	5,552.0
					Earth @ Stem Transitions	=	73.3	3.83	281.1
Resisting/Overturning Ratio		=	1.49		Footing Weight	=	1,512.5	2.75	4,159.4
Vertical Loads used for Soil Pressure =		5,952.2	lbs		Key Weight	=	200.0	3.50	700.0
					Vert. Component	=			
					Total =		5,952.2 lbs	R.M.=	21,364.2

* Axial live load NOT included in total displayed, or used for overturning
resistance, but is included for soil pressure calculation.

Vertical component of active lateral soil pressure IS NOT considered in
the calculation of Sliding Resistance.

Vertical component of active lateral soil pressure IS NOT considered in
the calculation of Overturning Resistance.

Tilt

Horizontal Deflection at Top of Wall due to settlement of soil

(Deflection due to wall bending not considered)

Soil Spring Reaction Modulus	250.0 pci
Horizontal Defl @ Top of Wall (approximate only)	0.271 in

The above calculation is not valid if the heel soil bearing pressure exceeds that of the toe,
because the wall would then tend to rotate into the retained soil.

DESIGN EXAMPLE 3   **Page 1 of 6**

Design Data

    Code: IBC '15

    Eq. Fluid Press = 30 pcf

    Soil Bearing = 3000 psf

    Soil Density = 110 pcf

    Passive = 300 pcf

    Surcharge / Axial as shown

    f_c' =   2500 psi

    f_y =   60,000 psi

    Neglect soil over toe for passive

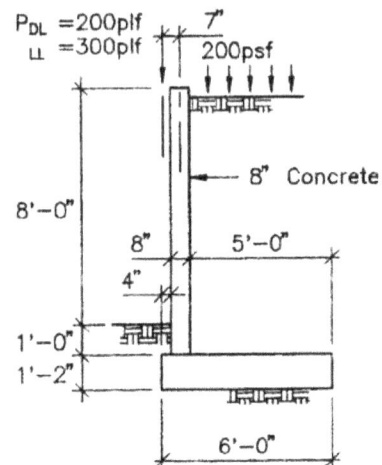

Stem @ Base (= +0.0)

$$M_u = \left[\frac{30 \times 9^2}{2} \times \frac{9}{3} \times 1.6\right] + \left[200 \times \frac{30}{110} \times \frac{9^2}{2} \times 1.6\right] + \left[200 \times 1.2 + 300 \times 1.6 \frac{7}{12}\right] = 9{,}787 \,^{\text{'\#}}$$

$$d = 8 - 2 - \frac{.75}{2} = 5.63''$$

The general solution for A_s (per CRSI)

$$= \frac{1.7\, f_c'\, bd}{2\, f_y} - \frac{1}{2}\sqrt{\frac{2.89\,(f_c'\, bd)^2}{f_y^{\,2}} - \frac{6.8\, f_c'\, b M_u}{\phi\, f_y^{\,2}}}$$

For b (unit stem width) = 12", f_y = 60 ksi: this reduces to:

$$A_s = 0.17\, f_c'\, d - \sqrt{.029\,(f_c'\, d)^2 - .0063\, f_c'\, M_u} \quad [M_u \text{ in in-kips}]$$

$$= .17 \times 2.5 \times 5.63 - \sqrt{.029\,(2.5 \times 5.63)^2 - .0063 \times 2.5 \times 9.787 \times 12}$$

$$= 0.42 \text{ sq. in./ft.}$$

A_s (Per above CRSI) = 0.42 sq. in./ft.

$$\rho_{min} = \frac{200}{f_y} = .0033$$

or $1.33 \times 0.42 = 0.56$ (See ACI '14, 9.6.1.3)

    <u>Use # 6 @ 9"</u> $A_s = \frac{.44}{.75} = .59$ $\rho = \frac{.44}{9 \times 5.63} = .0087 > .0033$

$$a = \frac{.44 \times 60}{.85 \times 2.5 \times 9} = 1.38''$$

$$M_n = .59 \times 60,000 \left(5.63 - \frac{1.38}{2} \right) \times .90 \times \frac{1}{12} = 13,115^{\#} > 9,787 \quad OK$$

$$\phi\uparrow$$

$$\text{Stress ratio} = \frac{9787}{13,115} = 0.746$$

Check Stem Shear

$$V_u = \frac{30 \times 9^2}{2} \times 1.6 + 200 \times \frac{30}{110} \times 9 \times 1.6 = 2729^{\#}$$

$$v_u = \frac{2729}{12 \times 5.63} = 40.4 \text{ psi}$$

$$v_{allow} = \varphi \, 2\sqrt{f'_c} = .75 \times 2\sqrt{2500} = 75 \text{ psi} \quad OK$$

Check embedment into footing

$$\text{For hooked bar} = \frac{.02 \times f_y \, d_b \times 0.7}{\sqrt{f'_c}}$$

or $8d_b$ or 6"

$$= \frac{0.02 \times 60,000 \times .75 \times 0.7}{50} = 12.6"$$

Choose to reduce by stress level per ACI 25.4.10

$$\therefore \quad \text{embedment} = 12.6 \times .746 = 9.40"$$

Footing thickness required

$$= \quad 9.4" + 3" \text{ clear} = 12.4"$$

$$\underline{\text{use 14" thick}} \quad (d = 14 - 3 - .50 = 10.5")$$
$$\text{arbitrary} \uparrow$$

Lap length above footing

$$\ell d_b = \frac{0.024 \times 0.75 \times 60,000}{\sqrt{2500}} \times 0.746 = 16.1"$$

Assume Class B splice w/continuing. #6 bars above, then lap = 16.1 x 1.3 = 20.94".

Check Stem @ + 1.5 Above Footing

$$M_u = \frac{30 \times 7.5^2}{2} \times \frac{7.5}{3} \times 1.6 + 200 \frac{30}{110} \times \frac{7.5^2}{2} \times 1.6 + (200 \times 1.2 + 300 \times 1.6) \frac{7}{12}$$

$$= 6250''^{\#}$$

$A_s = 0.257$ [CRSI equation]   Use #6 @ 18" $A_S = \frac{.44}{1.5} = 0.293 > 0.257$

$\underline{M_n @ + 1.5'}$ $a = \frac{.44 \times 60}{.85 \times 2.5 \times 18} = 0.69$

$$M_n = \frac{.44}{1.5} \times 60{,}000 \times \left(5.63 - \frac{.69}{2}\right) \times .9 \times \frac{1}{12} = 6976'^{\#} > 6250 \quad \text{OK}$$

Lap length over dowels $= \frac{.024 \times .75 \times 60{,}000}{\sqrt{2500}} \times 1.3 = 28.1''$ (assuming Class B

      splice)

Extend ftg. dowels 30" high.

Check Shear

$$V_u = \frac{30 \times 7.5^2}{2} \times 1.6 + 200 \frac{30}{110} \times 7.5 \times 1.6 = 2005^{\#}$$

$$v = \frac{2005}{12 \times 5.63} = 29.7 < 75\ psi$$

DESIGN EXAMPLE 3   **Page 4 of 6**

Stability Check

Item	Wt.		arm		+M	-M
Stem $\frac{8}{12}$ x 150 x 9	900	x	.67	=	600	
Earth 5 x 110 x 9	4950	x	3.5	=	17325	
Surcharge 5 x 200	1000	x	3.5	=	3500	
Axial DL	200	x	$\frac{1}{12}$	=	17	
Axial LL	300	x	$\frac{1}{12}$	=	25	
Footing 6 x 1.17 x 150	1053	x	3.0	=	3159	
OTM earth = $\dfrac{30 \times 10.17^2}{2}$			$\dfrac{10.17}{3}$	=		5259
Surcharge = $200 \times \dfrac{30}{110} \times 10.17$			$\dfrac{10.17}{2}$	=		2821
	8403				24,626	8080

w/o axial LL = 8103

$\bar{x}$ (from front edge of footing) = $\dfrac{24,626 - 8080}{8403} = 1.97'$

$e = \dfrac{6}{2} - 1.97 = 1.03' = 12.36"$ Middle-third = $\dfrac{6}{6} = \dfrac{1}{3}\dfrac{6}{6} = 1.00$ ft. $= 12.0$ in.

$\therefore$ Slightly outside middle third

Soil ρ = $\dfrac{8403}{.75 \times 6 - 1.5 \times 1.03} = 2829$ psf

OTM ratio (w/o axial LL) = $\dfrac{24,626 - 25}{8080} = 3.04$

NOTE: If surcharge is live load it should be excluded from overturning and sliding resistance and lateral pressure reduced accordingly.

DESIGN EXAMPLE 3 Page 5 of 6

Check Sliding

Total lateral = $\dfrac{30 \; x \; 10.17^2}{2} + 200 \; x \; \dfrac{30}{110} \; x \; 10.17 = 2106$

Passive resistance

$= \dfrac{300 \; x \; 2.17^2}{2} - \dfrac{300 \; x \; 1^2}{2} = 556 \text{ lbs.}$

Friction resistance (w/o axial LL)

$= 8103 \; x \; 0.40 = 3241^{\#}$

Sliding factor of safety $= \dfrac{556 + 3241}{2106} = 1.80$

OK > 1.5

Check Heel

Factored total vert. load = 8403 x 1.2 + 300 x 1.6 = 10,204$^{\#}$

Soil p (using e = 1.02') $= \dfrac{10,204}{.75 \; x \; 6 \; - \; 1.5 \; x \; 1.02}$

$= 3456 \text{ psf} > 3,000 \text{ psf}$

Soil p @ $P_1 = \dfrac{4.94}{5.94} \; x \; 3456 = 2874 \text{ psf}$

$M_u\!\downarrow \; = \; (1.17 \; x \; 150 \; x \; 1.2 + 110 \; x \; 9 \; x \; 1.2 + 200 \; x \; 1.2) \; x \; \dfrac{5^2}{2} = 20,483^{'\#}$

$M_u\!\uparrow \; = \; \dfrac{2874 \; x \; 4.94}{2} \; x \; \dfrac{4.94}{3} = 11,689^{'\#}$

$M_{design} \; = \; 20,483 - 11,689 = 8794^{'\#}$

A_s (w/ d = 11.5") = 0.164 sq. in. $\qquad \rho_{min} = \dfrac{200}{f_y} = .0033$

A_s min required = .0033 x 12 x 11.5 = 0.455 sq. in. or 1.33 x .164 = 0.219

Use #5 e 9" to match stem dowel spacing.

$A_s \; = \; \dfrac{0.31}{.75} = 0.41 > 0.219$

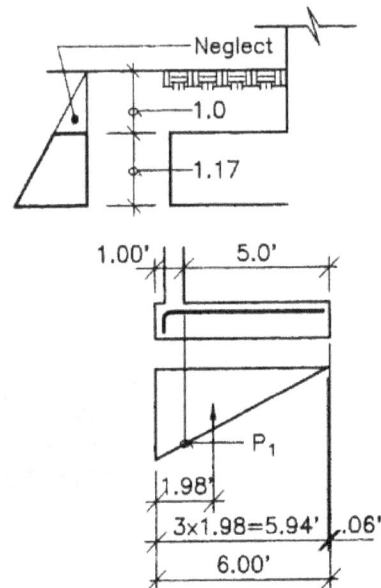

DESIGN EXAMPLE 3   **Page 6 of 6**

Check Shear @ Heel

V_u @ face of stem = (1.17 x 150 x 1.2) + (110 x 9 x 1.2) + (200 x 5 x 1.6)

$= 8593^{\#}$

$v = \dfrac{8593}{12 \ x \ 11.5} = 62.2$ psi $< 2\varphi\sqrt{f_c'} = 75$ OK

Extension of top bar from stem face

$= \dfrac{0.02 \ x \ 0.625 \ x \ 60{,}000 \ x \ 0.7}{\sqrt{2500}} = 10.5$ inches

DESIGN EXAMPLE 3 Report Printout

Basics of Retaining Wall Design

Title   EX-3
Job #   Dsgnr   HB   Date   2 JUL 2018
Description ..
EX-3

Page 1

This Wall in File: c:\Users\chris\Dropbox\RetainPro 10 Project Files\HB\Hugh. RPX

RetainPro (c) 1987-2018, Build 11.18.06.22
License : KW-06000216
License To : RetainPro Development Center

Cantilevered Retaining Wall

Code: IBC 2015,ACI 318-14,ACI 530-13

Criteria

Retained Height	=	9.00 ft
Wall height above soil	=	0.00 ft
Slope Behind Wall	=	0.00
Height of Soil over Toe	=	12.00 in
Water height over heel	=	0.0 ft

Soil Data

Allow Soil Bearing	=	3,000.0 psf
Equivalent Fluid Pressure Method		
Active Heel Pressure	=	30.0 psf/ft
Passive Pressure	=	300.0 psf/ft
Soil Density, Heel	=	110.00 pcf
Soil Density, Toe	=	100.00 pcf
Footing||Soil Friction	=	0.400
Soil height to ignore for passive pressure	=	12.00 in

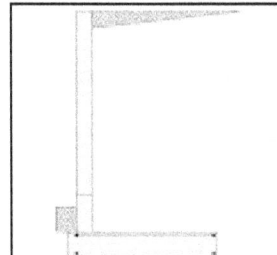

Surcharge Loads

Surcharge Over Heel	=	200.0 psf
Used To Resist Sliding & Overturning		
Surcharge Over Toe	=	0.0 psf
NOT Used for Sliding & Overturning		

Axial Load Applied to Stem

Axial Dead Load	=	200.0 lbs
Axial Live Load	=	300.0 lbs
Axial Load Eccentricity	=	7.0 in

Lateral Load Applied to Stem

Lateral Load	=	0.0 #/ft
...Height to Top	=	0.00 ft
...Height to Bottom	=	0.00 ft
Load Type	=	Wind (W) (Service Level)
Wind on Exposed Stem	=	0.0 psf
(Service Level)		
Wind acts left-to-right toward retention side.		

Adjacent Footing Load

Adjacent Footing Load	=	0.0 lbs
Footing Width	=	0.00 ft
Eccentricity	=	0.00 in
Wall to Ftg CL Dist	=	0.00 ft
Footing Type		Line Load
Base Above/Below Soil at Back of Wall	=	0.0 ft
Poisson's Ratio	=	0.300

Design Summary

Wall Stability Ratios

Overturning	=	3.04 OK
Sliding	=	1.81 OK
Total Bearing Load	=	8,423 lbs
...resultant ecc.	=	12.48 in
Soil Pressure @ Toe	=	2,873 psf OK
Soil Pressure @ Heel	=	0 psf OK
Allowable	=	3,000 psf
Soil Pressure Less Than Allowable		
ACI Factored @ Toe	=	4,022 psf
ACI Factored @ Heel	=	0 psf
Footing Shear @ Toe	=	1.5 psi OK
Footing Shear @ Heel	=	62.2 psi OK
Allowable	=	75.0 psi

Sliding Calcs

Lateral Sliding Force	=	2,105.0 lbs
less 100% Passive Force	= -	564.2 lbs
less 100% Friction Force	= -	3,249.3 lbs
Added Force Req'd	=	0.0 lbs OK
...for 1.5 Stability	=	0.0 lbs OK

Vertical component of active lateral soil pressure IS
NOT considered in the calculation of soil bearing.

Load Factors

Building Code	IBC 2015, ACI
Dead Load	1.200
Live Load	1.600
Earth, H	1.600
Wind, W	1.000
Seismic, E	1.000

Stem Construction

		2nd	Bottom	
		Stem OK	Stem OK	
Design Height Above Ftg	ft =	1.50	0.00	
Wall Material Above "Ht"	=	Concrete	Concrete	
Design Method	=	LRFD	LRFD	ASD
Thickness		8.00	8.00	
Rebar Size	=	# 6	# 6	
Rebar Spacing	=	18.00	9.00	
Rebar Placed at	=	Edge	Edge	

Design Data

fb/FB + fa/Fa	=	0.897	0.762

Total Force @ Section				
Service Level	lbs =			
Strength Level	lbs =	2,004.5	2,729.5	3,600.0

Moment...Actual				
Service Level	ft-# =			
Strength Level	ft-# =	6,249.5	9,786.5	12,000.0
Moment....Allowable	ft-# =	6,968.1	13,022.4	

Shear.....Actual				
Service Level	psi =			
Strength Level	psi =	29.7	40.4	31.4
Shear....Allowable	psi =	75.0	75.0	
Anet (Masonry)	in2 =			
Rebar Depth 'd'	in =	5.63	5.63	

Masonry Data

f'm	psi =	
Fs	psi =	
Solid Grouting	=	
Modular Ratio 'n'		
Wall Weight	psf = 100.0	100.0
Short Term Factor	=	
Equiv. Solid Thick.	=	
Masonry Block Type	=	Medium Weight
Masonry Design Method	=	ASD

Concrete Data

f'c	psi =	2,500.0	2,500.0
Fy	psi =	60,000.0	60,000.0

DESIGN EXAMPLE 3	Report Printout

Basics of Retaining Wall Design

Title EX-3
Job #   Dsgnr. HB   Date 2 JUL 2018
Description...
EX-3

Page 1

This Wall in File: c:\Users\chris\Dropbox\RetainPro 10 Project Files\HB\Hugh.RPX

RetainPro (c) 1987-2018, Build 11.18.08.22
License : KW-06000216
License To : RetainPro Development Center

Cantilevered Retaining Wall

Code: IBC 2015, ACI 318-14, ACI 530-13

Criteria

Retained Height	=	9.00 ft
Wall height above soil	=	0.00 ft
Slope Behind Wall	=	0.00
Height of Soil over Toe	=	12.00 in
Water height over heel	=	0.0 ft

Soil Data

Allow Soil Bearing	=	3,000.0 psf
Equivalent Fluid Pressure Method		
Active Heel Pressure	=	30.0 psf/ft
Passive Pressure	=	300.0 psf/ft
Soil Density, Heel	=	110.00 pcf
Soil Density, Toe	=	100.00 pcf
Footing||Soil Friction	=	0.400
Soil height to ignore for passive pressure	=	12.00 in

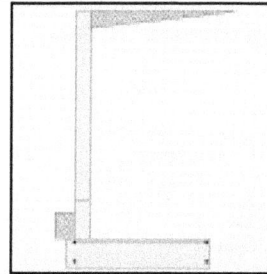

Surcharge Loads

Surcharge Over Heel	=	200.0 psf
Used To Resist Sliding & Overturning		
Surcharge Over Toe	=	0.0 psf
NOT Used for Sliding & Overturning		

Lateral Load Applied to Stem

Lateral Load	=	0.0 #/ft
.. Height to Top	=	0.00 ft
.. Height to Bottom	=	0.00 ft
Load Type	=	Wind (W)
		(Service Level)

Wind on Exposed Stem = 0.0 psf
(Service Level)
Wind acts left-to-right toward retention side.

Adjacent Footing Load

Adjacent Footing Load	=	0.0 lbs
Footing Width	=	0.00 ft
Eccentricity	=	0.00 in
Wall to Ftg CL Dist	=	0.00 ft
Footing Type		Line Load
Base Above/Below Soil at Back of Wall	=	0.0 ft
Poisson's Ratio	=	0.300

Axial Load Applied to Stem

Axial Dead Load	=	200.0 lbs
Axial Live Load	=	300.0 lbs
Axial Load Eccentricity	=	7.0 in

Design Summary

Wall Stability Ratios

Overturning	=	3.04 OK
Sliding	=	1.81 OK
Total Bearing Load	=	8,423 lbs
...resultant ecc.	=	12.48 in
Soil Pressure @ Toe	=	2,873 psf OK
Soil Pressure @ Heel	=	0 psf OK
Allowable	=	3,000 psf

Soil Pressure Less Than Allowable

ACI Factored @ Toe	=	4,022 psf
ACI Factored @ Heel	=	0 psf
Footing Shear @ Toe	=	1.5 psi OK
Footing Shear @ Heel	=	62.2 psi OK
Allowable	=	75.0 psi

Sliding Calcs

Lateral Sliding Force	=	2,105.0 lbs
less 100% Passive Force	= -	554.2 lbs
less 100% Friction Force	= -	3,249.3 lbs
Added Force Req'd	=	0.0 lbs OK
...for 1.5 Stability	=	0.0 lbs OK

Vertical component of active lateral soil pressure IS NOT considered in the calculation of soil bearing.

Load Factors

Building Code	IBC 2015,ACI
Dead Load	1.200
Live Load	1.600
Earth, H	1.600
Wind, W	1.000
Seismic, E	1.000

Stem Construction

		2nd	Bottom	
Design Height Above Ftg	ft =	Stem OK 1.50	Stem OK 0.00	
Wall Material Above "Ht"	=	Concrete	Concrete	
Design Method	=	LRFD	LRFD	ASD
Thickness	=	8.00	8.00	
Rebar Size	=	# 6	# 6	
Rebar Spacing	=	18.00	9.00	
Rebar Placed at	=	Edge	Edge	

Design Data

fb/FB + fa/Fa	=	0.897	0.752	
Total Force @ Section				
Service Level	lbs =			
Strength Level	lbs =	2,004.5	2,729.5	3,600.0
Moment...Actual				
Service Level	ft-# =			
Strength Level	ft-# =	6,249.5	9,786.5	12,000.0
Moment.....Allowable	ft-# =	6,968.1	13,022.4	
Shear.....Actual				
Service Level	psi =			
Strength Level	psi =	29.7	40.4	31.4
Shear.....Allowable	psi =	75.0	75.0	
Anet (Masonry)	in2 =			
Rebar Depth 'd'	in =	5.63	5.63	

Masonry Data

f'm	psi =			
Fs	psi =			
Solid Grouting	=			
Modular Ratio 'n'	=			
Wall Weight	psf =	100.0	100.0	
Short Term Factor	=			
Equiv. Solid Thick.	=			
Masonry Block Type	=	Medium Weight		
Masonry Design Method	=	ASD		

Concrete Data

f'c	psi =	2,500.0	2,500.0	
Fy	psi =	60,000.0	60,000.0	

DESIGN EXAMPLE 3

Report Printout

Basics of Retaining Wall Design

Title EX-3
Job #
Description
EX-3

Desgnr HB

Date 2 JUL 2018

Page 2

This Wall in File: c:\Users\chris\Dropbox\RetainPro 10 Project Files\HB\Hugh.RPX

RetainPro (c) 1987-2018, Build 11.18.08.22
License : KW-06006216
License To : RetainPro Development Center

Cantilevered Retaining Wall

Code: IBC 2015,ACI 318-14,ACI 530-13

Concrete Stem Rebar Area Details

2nd Stem	Vertical Reinforcing	Horizontal Reinforcing		
As (based on applied moment)	0.2617 in2/ft			
(4/3) * As	0.349 in2/ft	Min Stem T&S Reinf Area 1.440 in2		
200bd/fy : 200(12)(5.625)/60000	0.225 in2/ft	Min Stem T&S Reinf Area per ft of stem Height 0.192 in2/ft		
0.0018bh : 0.0018(12)(8)	0.1728 in2/ft	Horizontal Reinforcing Options		
	-------------		One layer of	Two layers of :
Required Area	0.2617 in2/ft		#4 @ 12.50 in	#4 @ 25.00 in
Provided Area	0.2933 in2/ft		#5 @ 19.38 in	#5 @ 38.75 in
Maximum Area	0.762 in2/ft		#6 @ 27.50 in	#6 @ 55.00 in

Bottom Stem	Vertical Reinforcing	Horizontal Reinforcing		
As (based on applied moment)	0.4099 in2/ft			
(4/3) * As	0.5465 in2/ft	Min Stem T&S Reinf Area 0.288 in2		
200bd/fy : 200(12)(5.625)/60000	0.225 in2/ft	Min Stem T&S Reinf Area per ft of stem Height 0.192 in2/ft		
0.0018bh : 0.0018(12)(8)	0.1728 in2/ft	Horizontal Reinforcing Options		
	-------------		One layer of	Two layers of :
Required Area	0.4099 in2/ft		#4 @ 12.50 in	#4 @ 25.00 in
Provided Area	0.5867 in2/ft		#5 @ 19.38 in	#5 @ 38.75 in
Maximum Area	0.762 in2/ft		#6 @ 27.50 in	#6 @ 55.00 in

Footing Dimensions & Strengths

Toe Width	=	0.33 ft
Heel Width	=	5.66
Total Footing Width	=	5.99
Footing Thickness	=	14.00 in
Key Width	=	3.00 in
Key Depth	=	0.00 in
Key Distance from Toe	=	0.00 ft

f'c = 2,500 psi   Fy = 60,000 psi
Footing Concrete Density = 150.00 pcf
Min. As % = 0.0018
Cover @ Top 2.00   @ Btm = 3.00 in

Footing Design Results

		Toe	Heel
Factored Pressure	=	4,022	0 psf
Mu' : Upward	=	215	0 ft-#
Mu' : Downward	=	19	21,418 ft-#
Mu : Design	=	196	21,418 ft-#
Actual 1-Way Shear	=	1.46	62.16 psi
Allow 1-Way Shear	=	40.00	75.00 psi
Toe Reinforcing	=	#6 @ 18.00 in	
Heel Reinforcing	=	#5 @ 18.00 in	
Key Reinforcing	=	None Spec'd	

Other Acceptable Sizes & Spacings

Toe   Not req'd: Mu < phi*5*lambda*sqrt(f'c)*Sm
Heel #4 @ 5.22 in, #5 @ 8.09 in, #6 @ 11.48 in, #7 @ 15.65 in, #8 @ 20.61 in, #9 @ 26
Key   No key defined

Min footing T&S reinf Area	1.81	in2
Min footing T&S reinf Area per foot	0.30	in2/ft

If one layer of horizontal bars	If two layers of horizontal bars
#4 @ 7.94 in	#4 @ 15.87 in
#5 @ 12.30 in	#5 @ 24.60 in
#6 @ 17.46 in	#6 @ 34.92 in

DESIGN EXAMPLE 3   Report Pr

Basics of Retaining Wall Design

Title EX-3
Job #   Dsgnr: HB   Date   Page   3
Description...   2 JUL 2016
EX-3

This Wall in File: c:\Users\chris\Dropbox\RetainPro 10 Project Files\HB\Hugh.RPX

RetainPro (c) 1987-2016 Build 11.16.06.22
License : KW-06000016
License To : RetainPro Development Center

Cantilevered Retaining Wall   Code: IBC 2015, ACI 318-14, ACI 530-13

Summary of Overturning & Resisting Forces & Moments

Item	....OVERTURNING....				....RESISTING....		
	Force lbs	Distance ft	Moment ft-#		Force lbs	Distance ft	Moment ft-#
Heel Active Pressure =	1,550.4	3.30	5,254.2	Soil Over Heel =	4,943.4	3.49	17,268.9
Surcharge over Heel =	554.5	5.08	2,818.9	Sloped Soil Over Heel =			
Surcharge Over Toe =				Surcharge Over Heel =	998.7	3.49	3,488.7
Adjacent Footing Load =				Adjacent Footing Load =			
Added Lateral Load =				Axial Dead Load on Stem =	200.0	0.08	16.0
Load @ Stem Above Soil =				* Axial Live Load on Stem =	300.0	0.08	24.0
=				Soil Over Toe =	33.0	0.17	5.4
				Surcharge Over Toe =			
Total	2,105.0	O.T.M.	8,073.1	Stem Weight(s) =	900.0	0.66	597.0
=				Earth @ Stem Transitions =			
Resisting/Overturning Ratio		=	3.04	Footing Weight =	1,048.3	3.00	3,139.5
Vertical Loads used for Soil Pressure =	8,423.3	lbs		Key Weight =			
				Vert. Component =			
				Total =	8,123.3 lbs R.M.=		24,515.6

* Axial live load NOT included in total displayed, or used for overturning
resistance, but is included for soil pressure calculation.

Vertical component of active lateral soil pressure IS NOT considered in
the calculation of Sliding Resistance.

Vertical component of active lateral soil pressure IS NOT considered in
the calculation of Overturning Resistance.

Tilt

Horizontal Deflection at Top of Wall due to settlement of soil

(Deflection due to wall bending not considered)

Soil Spring Reaction Modulus	250.0	pci
Horizontal Defl @ Top of Wall (approximate only)	0.120	in

The above calculation is not valid if the heel soil bearing pressure exceeds that of the toe
because the wall would then tend to rotate into the retained soil.

DESIGN EXAMPLE 4   **Page 1 of 4**

Design Data

    Code: IBC '15

    Equivalent fluid press. = 30 pcf

    Wind on fence -= 15 psf

    Soil bearing = 1500 psf

    Passive = 350 pcf

    Soil density = 110 pcf

    μ = 0.45 = Footing friction coeff.

    $f_m' = 1500$

    Use LRFD method

    $f_y = 60,000$

    $f_c' = 2500$

    $f_y' = 60,000$

Neglect soil over toe for passive

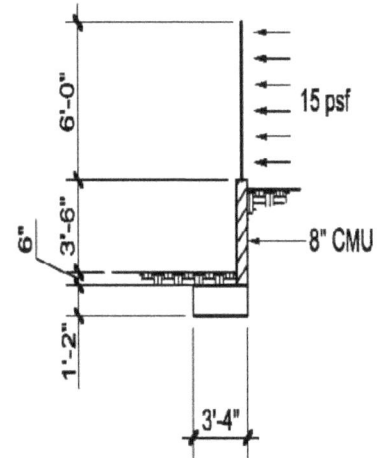

Check Fence

$$M \text{ @ base} = \frac{15 \times 6^2}{2} = 270^{'\#} \quad \text{(for design of fence connection to wall)}$$

    Lateral @ bott. of fence = 15 x 6 = 90 plf

Check stem @ base (Ht = 0.0)

    Use LRFD Design Method

    (Load factor wind and earth pressure = 1.6)

$$M_u = \left[15 \times 6 \left(\frac{6}{2} + 4 \right) + \frac{30 \times 4^2}{2} \times \frac{4}{3} \right] \times 1.6 = 1520 \text{ ft-lbs.}$$

    Use 8" block solid grouted

    $f_b = 1500 \times .33 = 500$ psi

    Try #5 @ 24 in. ($A_s = 0.31/2.0 = 0.155$)

    d = 5.25"

DESIGN EXAMPLE 4

$$a = \frac{0.155 \times 60{,}000}{0.80 \times 1500 \times 0.80 \times 12} \cong 0.81$$

$\phi M_n = 0.90 \times 0.155 \times 60{,}000\ [5.25 - (0.81/2)] \times \underline{1/12} = 3379$ ft-lbs.

Stress ratio $= \dfrac{1520}{3379} = 0.45$

Stem Shear

Lateral @ base $= \left(15\ psf \times 6' + \dfrac{30\ x\ 4^2}{2}\right) \times 1.6 = 528$ lbs.

$v_u = \dfrac{528}{12\ x\ 5.25} = 8.38$ psi $< \sqrt{1500} = 38.7$ OK

$\phi V_u \times 0.80\ (4.0 - 1.75)\ \sqrt{1500} = 69.7$ psi

Embedment into ftg. w/ std. hook

$= \dfrac{.02 \times 60{,}000 \times 0.625 \times 0.7}{\sqrt{2500}} = 10.5"$ Use 6" min.

Min. ftg. Thickness $= 10.5 + 3.0 = 13.5"$ Use 14"

DESIGN EXAMPLE 4   **Page 3 of 4**

<u>Stability</u>

Item	Wt.	arm	+M	-M
Fence	-0-			
8" stem 4 x 78 psf	312	3.0	936	
Earth over toe 0.5 x 110 x 2.67	147	1.33	195	
Footing 1.17 x 150 x 3.33	584	1.67	975	
Wind OTM 15 x 6′		8.17		735
Soil OTM $= \dfrac{30 \ x \ 5.17^2}{2}$		1.72		690
	1043		2106	1425

$$\bar{x} = \frac{2106 - 1425}{1043} = 0.66 \qquad e = \frac{3.33}{2} - .66 = 1.0' = 12"$$

Middle third $e = \dfrac{3.33}{6} = .55$ $\therefore$ outside middle third

Soil $p = \dfrac{1043}{.75 \ x \ 3.33 - 1.5 \ x \ 1.0} = 1043$ psf

Overturning ratio $= \dfrac{2106}{1425}$ x 1.48   (consider OK because of wind load)

<u>Check Sliding</u>

Total lateral $= 6 \ x \ 15 \ psf + \dfrac{30 \ x \ 5.17^2}{2} = 490$ lbs.

Friction resistance $= 1043$ x .45 $= 469$ lbs.

DESIGN EXAMPLE 4   **Page 4 of 4**

Passive resistance $= \dfrac{350 \ x \ 1.67^2}{2} - \dfrac{350 \ x \ .5^2}{2}$

$= 444$ lbs.

Sliding ratio $= \dfrac{444 + 469}{490} = 1.86 > 1.5$ OK

Critical for M

Critical for V

8.5"

1346

8" 1'–4" 8" 8"

3'–4"

Check Toe

Total factored vertical load $= 1067 \ x \ 1.2 = 1280$ lbs.

Factored soil $p = \dfrac{1280}{.75 \ x \ 3.33 - 1.5 \ x \ 1} = 1280$ psf

$M_u\uparrow \ = \ \dfrac{1280 \ x \ 2}{2} (3.33 - .67 - .50) = 2765$ ft-lbs.

$M_u\downarrow \ = \ (150 \ x \ 1.17 + 110 \ x \ .5) \ 2.67 \ x \ 1.33 \ x \ 1.2 = 982$ ft-lbs.

$M_{design} = \ 2765 - 982 = 1783$ ft-lbs.

$V_u \ = \ 1280 \ x \ 2 \ x \ \frac{1}{2} - (150 \ x \ 1.17 + 110 \ x \ .5) \ 2.67 \ x \ 1.2 = 543$ lbs.

$v_u \ = \ \dfrac{543}{12 \ x \ 10.5} = 4.3$ $v_{allow} = 76$ OK

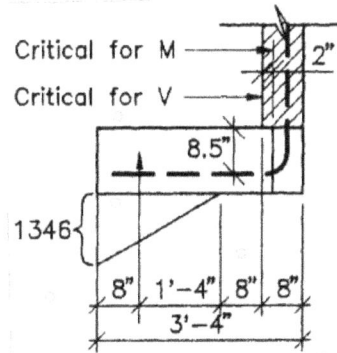

Toe Reinforcing

$M_u \ = \ 982$ ft-lbs.

$d \ = \ 14 - 3 - .5 = 10.5$ in.

A_s required $\ = \ 0.013$ A_s min $= \dfrac{200}{f_y} = .0033$

Min. reinf. $= \ .0033 \ x \ 12 \ x \ 10.5 = 0.416$

or $= \ 1.33 \ x \ .042 = 0.0559$

or $= \ .0018 \ x \ 12 \ x \ 10.5 = 0.227$

Select #5 @ 16   (to match stem dowel bars bent to toe)

$A_s = \dfrac{0.31}{1.33} = 0.23 > 0.227$ OK

DESIGN EXAMPLE 4   Report Printout

Basics of Retaining Wall Design

Title   EX-4
Job #
Description
EX-4

Dsgnr: HB   Date: 2 JUL 2018   Page: 1

This Wall in File: c:\Users\chris\Dropbox\RetainPro 10 Project Files\HB\Hugh.RPX

RetainPro (c) 1987-2018, Build 11.18.06.22
License : KW-06000215
License To : RetainPro Development Center

Cantilevered Retaining Wall

Code: IBC 2015, ACI 318-14, ACI 530-13

Criteria

Retained Height	=	4.00 ft
Wall height above soil	=	6.00 ft
Slope Behind Wall	=	0.00
Height of Soil over Toe	=	6.00 in
Water height over heel	=	0.0 ft

Soil Data

Allow Soil Bearing	=	1,500.0 psf
Equivalent Fluid Pressure Method		
Active Heel Pressure	=	30.0 psf/ft
Passive Pressure	=	350.0 psf/ft
Soil Density, Heel	=	110.00 pcf
Soil Density, Toe	=	110.00 pcf
Footing||Soil Friction	=	0.450
Soil height to ignore for passive pressure	=	6.00 in

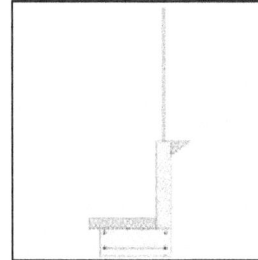

Surcharge Loads

Surcharge Over Heel	=	0.0 psf
Used To Resist Sliding & Overturning		
Surcharge Over Toe	=	0.0 psf
Used for Sliding & Overturning		

Axial Load Applied to Stem

Axial Dead Load	=	0.0 lbs
Axial Live Load	=	0.0 lbs
Axial Load Eccentricity	=	0.0 in

Lateral Load Applied to Stem

Lateral Load	=	0.0 #/ft
...Height to Top	=	0.00 ft
...Height to Bottom	=	0.00 ft
Load Type	=	Wind (W) (Service Level)
Wind on Exposed Stem (Service Level)	=	15.0 psf

Adjacent Footing Load

Adjacent Footing Load	=	0.0 lbs
Footing Width	=	0.00 ft
Eccentricity	=	0.00 in
Wall to Ftg CL Dist	=	0.00 ft
Footing Type		Line Load
Base Above/Below Soil at Back of Wall	=	0.0 ft
Poisson's Ratio	=	0.300

Design Summary

Wall Stability Ratios

Overturning	=	2.09 OK
Sliding	=	2.13 OK
Total Bearing Load	=	1,336 lbs
...resultant ecc	=	5.99 in
Soil Pressure @ Toe	=	762 psf OK
Soil Pressure @ Heel	=	40 psf OK
Allowable	=	1,500 psf
Soil Pressure Less Than Allowable		
ACI Factored @ Toe	=	1,067 psf
ACI Factored @ Heel	=	56 psf
Footing Shear @ Toe	=	6.7 psi OK
Footing Shear @ Heel	=	2.7 psi OK
Allowable	=	75.0 psi

Sliding Calcs

Lateral Sliding Force	=	490.4 lbs
less 100% Passive Force =	-	442.4 lbs
less 100% Friction Force =	-	601.1 lbs
Added Force Req'd	=	0.0 lbs OK
...for 1.5 Stability	=	0.0 lbs OK

Vertical component of active lateral soil pressure IS NOT considered in the calculation of soil bearing

Load Factors

Building Code	IBC 2015, ACI
Dead Load	1.200
Live Load	1.600
Earth, H	1.600
Wind, W	1.600
Seismic, E	1.000

Stem Construction

		2nd	Bottom
			Stem OK
Design Height Above Ftg ft =		4.00	0.00
Wall Material Above "Ht"	=	Fence	Masonry
Design Method	=	LRFD	LRFD ASD
Thickness	=		8.00
Rebar Size	=		# 5
Rebar Spacing	=		24.00
Rebar Placed at	=		Edge

Design Data

fb/FB + fa/Fa	=		0.446
Total Force @ Section			
Service Level	lbs =		
Strength Level	lbs =	90.0	528.0
Moment....Actual			
Service Level	ft-# =		
Strength Level	ft-# =	270.0	1,520.0
Moment....Allowable	ft-# =		3,411.6
Shear.....Actual			
Service Level	psi =		
Strength Level	psi =		5.8
Shear....Allowable	psi =		69.7
Anet (Masonry)	in2 =		91.50
Rebar Depth 'd'	in =		5.25

Masonry Data

f'm	psi =	1,500
Fy	psi =	60,000
Solid Grouting		Yes
Modular Ratio 'n'	=	21.48
Wall Weight	psf =	78.0
Equiv. Solid Thick.	=	7.60
Masonry Block Type	=	Medium Weight
Masonry Design Method	=	LRFD

Concrete Data

f'c	psi =	
Fy	psi =	

DESIGN EXAMPLE 4   Report Printout

Basics of Retaining Wall Design

Title EX-4
Job #
Description
EX-4

Dsgnr: HB

Date: 2 JUL 2018

Page: 2

This Wall in File: c:\Users\chris\Dropbox\RetainPro 10 Project Files\HB\Hugh.RPX

RetainPro (c) 1987-2018, Build 11.18.06.22
License : KW-06006215
License To : RetainPro Development Center

Cantilevered Retaining Wall

Code: IBC 2015,ACI 318-14,ACI 530-13

Footing Dimensions & Strengths

Toe Width	=	2.66 ft
Heel Width	=	0.67
Total Footing Width	=	3.33
Footing Thickness	=	14.00 in
Key Width	=	0.00 in
Key Depth	=	0.00 in
Key Distance from Toe	=	0.00 ft

$f'c$ = 2,500 psi Fy = 60,000 psi
Footing Concrete Density = 150.00 pcf
Min. As % = 0.0018
Cover @ Top   2.00   @ Btm = 3.00 in

Footing Design Results

		Toe	Heel
Factored Pressure	=	1,067	96 psf
Mu' : Upward	=	2,823	28 ft-#
Mu' : Downward	=	1,099	248 ft-#
Mu : Design	=	1,724	220 ft-#
Actual 1-Way Shear	=	6.75	2.70 psi
Allow 1-Way Shear	=	40.00	40.00 psi
Toe Reinforcing	=	#5 @ 16.00 in	
Heel Reinforcing	=	None Spec'd	
Key Reinforcing	=	None Spec'd	

Other Acceptable Sizes & Spacings

Toe: Not req'd: Mu < phi*5*lambda*sqrt(f'c)*Sm
Heel: Not req'd: Mu < phi*5*lambda*sqrt(f'c)*Sm
Key: No key defined

Min footing T&S reinf Area   1.01 in2
Min footing T&S reinf Area per foot   0.30 in2 ft

If one layer of horizontal bars:	If two layers of horizontal bars:
#4@ 7.94 in	#4@ 15.87 in
#5@ 12.30 in	#5@ 24.60 in
#6@ 17.46 in	#6@ 34.92 in

Summary of Overturning & Resisting Forces & Moments

		OVERTURNING					RESISTING		
Item		Force lbs	Distance ft	Moment ft-#			Force lbs	Distance ft	Moment ft-#
Heel Active Pressure	=	400.4	1.72	680.6	Soil Over Heel	=	294.8	3.00	882.9
Surcharge over Heel	=				Sloped Soil Over Heel	=			
Surcharge Over Toe	=				Surcharge Over Heel	=			
Adjacent Footing Load	=				Adjacent Footing Load	=			
Added Lateral Load	=				Axial Dead Load on Stem	=			
Load @ Stem Above Soil	=	90.0	8.17	735.0	* Axial Live Load on Stem	=			
	=				Soil Over Toe	=	146.3	1.33	194.6
					Surcharge Over Toe	=			
Total		490.4	O.T.M.	1,424.6	Stem Weight(s)	=	312.0	2.99	933.9
	=				Earth @ Stem Transitions	=			
Resisting/Overturning Ratio	=	2.09			Footing Weight	=	582.8	1.67	970.3
Vertical Loads used for Soil Pressure =		1,335.9 lbs			Key Weight	=			
					Vert. Component	=			
						Total =	1,335.9 lbs	R.M.=	2,981.7

* Axial live load NOT included in total displayed, or used for overturning resistance, but is included for soil pressure calculation.

Vertical component of active lateral soil pressure IS NOT considered in the calculation of Sliding Resistance.

Vertical component of active lateral soil pressure IS NOT considered in the calculation of Overturning Resistance.

Tilt

Horizontal Deflection at Top of Wall due to settlement of soil

(Deflection due to wall bending not considered)

Soil Spring Reaction Modulus   250.0 pci
Horizontal Defl @ Top of Wall (approximate only)   0.064 in

The above calculation is not valid if the heel soil bearing pressure exceeds that of the toe, because the wall would then tend to rotate into the retained soil.

DESIGN EXAMPLE 5	**Page 1 of 3**

<u>Freestanding Yard Wall</u>

<u>Design Data</u>

    Code: IBC '15

    Assume California so seismic governs over
       wind.

    Soil bearing = 2000 psf

    Use 6" CMU solid grout

    Equiv. fluid pressure = 30 pcf

    Passive = 400 pcf

    μ = 0.40

    $f_m' = 1500$ $F_s = 24{,}000$

    $f_c' = 2000$ $f_y = 60{,}000$

Determine seismic force factor F_P/W_P

Per ASCE 7-05, 15.4.2 (Rigid nonbuilding
    structures)

From Hazard Maps for short period assume $S_s = 1.10$

Seismic Design Category D

$F_a = 1.0$

$S_{MS} = F_a\,S_s = 1.0 \times 1.10 = 1.10$

$S_{DS} = 0.67\,(1.10) = 0.73$ (Eq. 16-38)

<u>Equation (15.4.5), rewritten</u>

$F_p/W_p = 0.30\,S_{DS}\,I = 0.30 \times 0.73 \times 1.0 = 0.22$ (Assume I = 1.0)

Input $F_p/W_p = 0.22 \times \dfrac{1}{1.4} = 0.16$ for masonry ASD and overturning

Convert to ASD $\uparrow$

One-third stress increase permitted per IBC '15, 1807.3.2.1

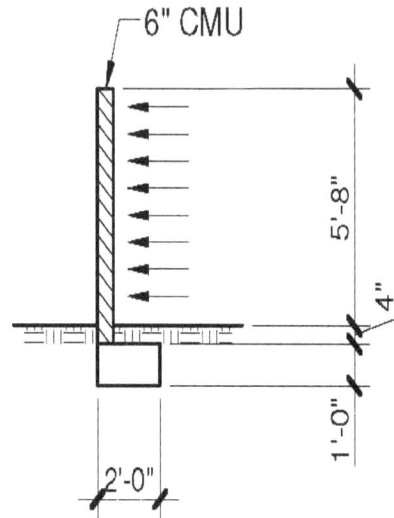

DESIGN EXAMPLE 5   Page 2 of 3

Stem Design, Masonry

$$M \quad = \quad (0.16 \times 58 \text{ psf}) \; \frac{6^2}{2} = 167 \text{ ft-lbs.}$$

Use #4 e 48" @ center (d = 2.75")

$$A_s = \frac{.20}{4} = .05 \text{ sq in/ft}$$

$$np \quad = \quad \frac{21.5 \; x \; .20}{48 \; x \; 2.75} = .033 \qquad \frac{2}{kj} = 9.57 \qquad\qquad j = .92$$

$$M_s \quad = \quad 24{,}000 \times 1.33 \times .92 \times 2.75 \times .05 \times \frac{1}{12} = 338^{'\#} > 167 \quad \text{OK}$$

$$M_m \quad = \quad \frac{1500 \; x \; .33 \; x \; 12 \; x \; 2.75^2}{9.57} \; x \; 1.33 \; x \; \frac{1}{12} = 525^{'\#}$$

$$V \quad = \quad 0.16 \times 58 \times 6 = 55.7 \text{ lbs.}$$

$$\nu \quad = \quad \frac{55.7}{12 \; x \; 2.75} = 1.68 \text{ psi} \qquad\qquad \text{OK}$$

$$v \text{ (Allow)} = \sqrt{1500} = 38.7 > 1.69 \quad \text{OK}$$

Check Stability

$$\text{Overturning} = 6.0 \times .16 \times 58 \left(\frac{6}{2} + 1 \right) + \frac{30 \; x \; 1.33^2}{2} \; x \; \frac{1.33}{3} = 234 \text{ ft- lbs.}$$

$$\text{Resisting moment} = 6 \times 58 \text{ psf} \times .25 + 110 \times .33 \times 1.5 \times 1.25 + 1.0 \times 150 \times 2.0 \times 1.0$$

$$= \quad 455^{'\#}$$

$$W \quad = \quad 6 \quad \text{x } 58 + .33 \quad \text{x } 110 \times 1.5 + 2.0 \text{ x } 1 \times 150$$

$$= \quad 702 \text{ lbs.}$$

$$\bar{x} \quad = \quad \frac{455 - 234}{6 \; \text{x } \; 58 \; + \; .33 \; \text{x } \; 110 \; \text{x } \; 1.5 \; + \; 2.0 \; \text{x } \; 1 \; \text{x } \; 150} = 0.32$$

$$e \quad = \quad \frac{2.0}{2} - 0.32 = 0.68 \qquad \text{middle third} = \frac{2.0}{6} = .0.33$$

$$\therefore \qquad \text{outside middle third}$$

DESIGN EXAMPLE 5   **Page 3 of 3**

Soil p= $\dfrac{702}{.75 \; x \; 2.0 - 1.5 \; x \; .68}$ = 1463 psf < 2000 psf OK

Overturning ratio = $\dfrac{455}{234}$ = 1.94   OK w/seismic

Lateral force @ base of ftg. = .16 x 58 x 6 + (30 x 1.33^2)/2 = 82.2 lbs.

Sliding ratio = $\dfrac{702 \; x \; .40 + \left(\dfrac{400 \; x \; 1.33^2}{2} - \dfrac{400 \; x \; .33^2}{2} \right)}{82.2}$ = 7.45

<u>Check Heel Reinforcing</u>

Neglect upward soil p

$M_u \downarrow$ $\quad = \quad$ (1.0 x 150 + .33 x 110) $\dfrac{1.5^2}{2}$ x 1.2 = 252$^{\prime \#}$

M_u stem $\; = \;$ 167 x 1.6 = 267 ft-lbs. $\leftarrow$ governs

d $\qquad = \quad$ 12 - 3 - .5 = 8.5"

A_s req'd $\; = \;$ .04 $^{sq. \, in.}\!/_{ft.}$ (per CRSI equation)

$\rho_{min} \quad = \quad \dfrac{200}{f_y}$ = .0033

$\qquad\qquad\qquad\qquad\qquad\qquad$ $\downarrow$ Stem As

∴ A_s min = .0033 x 12 x 8.5 = .34 or 1.33 x .05 = .067

Use #4 e 48 (to match stem dowels) $\quad A_s = \dfrac{.20}{4}$ = .05 < .067 Consider OK

V_u $\qquad = \quad$ (1.0 x 150 x 1.5 + .33 x 110 x 1.5) 1.2 = 335$^{\#}$/ ft

v_u $\qquad = \quad \dfrac{335}{12 \; x \; 8.5}$ = 3.3 < 76 OK

DESIGN EXAMPLE 5   **Report Printout**

Basics of Retaining Wall Design

Title	EX-5	Page 1
Job #	Dsgnr HB	Date 2 JUL 2018
Description		
EX-5		

This Wall in File: c:\Users\chris\Dropbox\RetainPro 10 Project Files\HB\Hugh.RPX

RetainPro (c) 1987-2018, Build 11.18.06.22
License : KW-06006216
License To : RetainPro Development Center

Cantilevered Retaining Wall   Code: IBC 2015, ACI 318-14, ACI 530-13

Criteria

Retained Height	=	0.33 ft
Wall height above soil	=	5.67 ft
Slope Behind Wall	=	0.00
Height of Soil over Toe	=	4.00 in
Water height over heel	=	0.0 ft

Soil Data

Allow Soil Bearing	=	2,000.0 psf
Equivalent Fluid Pressure Method		
Active Heel Pressure	=	30.0 psf/ft
	=	
Passive Pressure	=	400.0 psf/ft
Soil Density, Heel	=	110.00 pcf
Soil Density, Toe	=	100.00 pcf
Footing||Soil Friction	=	0.400
Soil height to ignore for passive pressure	=	4.00 in

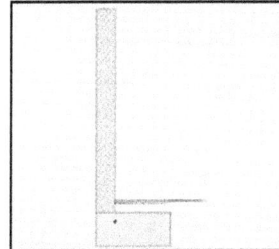

Surcharge Loads

Surcharge Over Heel	=	0.0 psf
Used To Resist Sliding & Overturning		
Surcharge Over Toe	=	0.0
Used for Sliding & Overturning		

Axial Load Applied to Stem

Axial Dead Load	=	0.0 lbs
Axial Live Load	=	0.0 lbs
Axial Load Eccentricity	=	0.0 in

Lateral Load Applied to Stem

Lateral Load	=	0.0 #/ft
Height to Top	=	0.00 ft
Height to Bottom	=	0.00 ft
Load Type	=	Wind (W) (Service Level)
Wind on Exposed Stem (Service Level)	=	0.0 psf

Stem Weight Seismic Load

F$_p$ / W$_p$ Weight Multiplier	=	0.220 g

Adjacent Footing Load

Adjacent Footing Load	=	0.0 lbs
Footing Width	=	0.00 ft
Eccentricity	=	0.00 in
Wall to Ftg CL Dist	=	0.00 ft
Footing Type		Line Load
Base Above/Below Soil at Back of Wall	=	0.0 ft
Poisson's Ratio	=	0.300
Added seismic base force		53.6 lbs

Design Summary

Wall Stability Ratios

Overturning	=	2.01 OK
Sliding	=	7.67 OK
Total Bearing Load	=	702 lbs
resultant ecc	=	8.09 in
Soil Pressure @ Toe	=	1,437 psf OK
Soil Pressure @ Heel	=	0 psf OK
Allowable	=	2,000 psf
Soil Pressure Less Than Allowable		
ACI Factored @ Toe	=	2,012 psf
ACI Factored @ Heel	=	0 psf
Footing Shear @ Toe	=	8.1 psi OK
Footing Shear @ Heel	=	2.8 psi OK
Allowable	=	67.1 psi

Sliding Calcs

Lateral Sliding Force	=	80.1 lbs
less 100% Passive Force =	-	333.3 lbs
less 100% Friction Force =	-	281.0 lbs
Added Force Req'd	=	0.0 lbs OK
for 1.5 Stability	=	0.0 lbs OK

Vertical component of active lateral soil pressure IS
NOT considered in the calculation of soil bearing.

Load Factors

Building Code	IBC 2015, ACI
Dead Load	1.200
Live Load	1.600
Earth, H	1.600
Wind, W	1.000
Seismic, E	1.000

Stem Construction

		Bottom
		Stem OK
Design Height Above Ftg	ft =	0.00
Wall Material Above "Ht"	=	Masonry
Design Method	=	ASD
Thickness	=	8.00
Rebar Size	=	# 4
Rebar Spacing	=	48.00
Rebar Placed at	=	Center

Design Data

fb/FB + fa/Fa	=	0.475

Total Force @ Section		
Service Level	lbs =	55.2
Strength Level	lbs =	90.0

Moment....Actual		
Service Level	ft-# =	161.0
Strength Level	ft-# =	270.0
Moment.....Allowable	=	339.2

Shear.....Actual		
Service Level	psi =	0.8
Strength Level	psi =	
Shear....Allowable	psi =	45.1
Anet (Masonry)	in2 =	67.50
Rebar Depth 'd'	in =	2.75

Masonry Data

f'm	psi =	1,500
Fs	psi =	32,000
Solid Grouting		Yes
Modular Ratio 'n'	=	21.48
Wall Weight	psf =	58.0
Short Term Factor	=	1.000
Equiv. Solid Thick.	in =	5.60
Masonry Block Type	=	Medium Weight
Masonry Design Method	=	ASD

Concrete Data

f'c	psi =	
Fy	psi =	

DESIGN EXAMPLE 5   Report Printout

Basics of Retaining Wall Design	Title EX-5	Dsgnr HB	Page 2
	Job #		Date 2 JUL 2018
	Description :		
	EX-5		

This Wall in File: c:\Users\chris\Dropbox\RetainPro 10 Project Files\HB\Hugh.RPX

RetainPro (c) 1987-2018, Build 11.18.08.22
License : KW-06000216
License To : RetainPro Development Center

Cantilevered Retaining Wall   Code: IBC 2015, ACI 318-14, ACI 530-13

Footing Dimensions & Strengths

Toe Width	=	0.00 ft
Heel Width	=	2.00
Total Footing Width	=	2.00
Footing Thickness	=	12.00 in
Key Width	=	0.00 in
Key Depth	=	0.00 in
Key Distance from Toe	=	0.00 ft

Fc = 2,000 psi   Fy = 60,000 psi
Footing Concrete Density = 150.00 pcf
Min. As % = 0.0018
Cover @ Top 2.00 @ Btm = 3.00 in

Footing Design Results

		Toe	Heel
Factored Pressure	=	2,012	0 psf
Mu' : Upward	=	0	0 ft-#
Mu' : Downward	=	0	293 ft-#
Mu : Design	=	0	293 ft-#
Actual 1-Way Shear	=	8.10	2.79 psi
Allow 1-Way Shear	=	35.78	35.78 psi
Toe Reinforcing	=	None Spec'd	
Heel Reinforcing	=	None Spec'd	
Key Reinforcing	=	None Spec'd	

Other Acceptable Sizes & Spacings

Toe: Not req'd: Mu < phi*5*lambda*sqrt(f'c)*Sm
Heel: Not req'd: Mu < phi*5*lambda*sqrt(f'c)*Sm
Key: No key defined

Min footing T&S reinf Area	0.52 in2
Min footing T&S reinf Area per foot	0.26 in2 /ft

If one layer of horizontal bars	If two layers of horizontal bars
#4@ 9.26 in	#4@ 18.52 in
#5@ 14.35 in	#5@ 28.70 in
#6@ 20.37 in	#6@ 40.74 in

Summary of Overturning & Resisting Forces & Moments

Item		....OVERTURNING... Force lbs	Distance ft	Moment ft-#			RESISTING.... Force lbs	Distance ft	Moment ft-#
Heel Active Pressure	=	26.5	0.44	11.8	Soil Over Heel	=	54.5	1.25	68.1
Surcharge over Heel	=				Sloped Soil Over Heel	=			
Surcharge Over Toe	=				Surcharge Over Heel	=			
Adjacent Footing Load	=				Adjacent Footing Load	=			
Added Lateral Load	=				Axial Dead Load on Stem	=			
Load @ Stem Above Soil	=				* Axial Live Load on Stem	=			
	=				Soil Over Toe	=			
Seismic Stem Self Wt		53.6	4.00	214.4	Surcharge Over Toe	=			
					Stem Weight(s)	=	348.0	0.25	87.0
Total		80.1	O.T.M.	226.1	Earth @ Stem Transitions	=			
					Footing Weight	=	300.0	1.00	300.0
Resisting/Overturning Ratio	=	2.01			Key Weight	=			
Vertical Loads used for Soil Pressure =		702.5 lbs			Vert. Component	=			
					Total =		702.5 lbs	R.M.=	455.1

If seismic is included, the OTM and sliding ratios may be 1.1 per section 1807.2.3 of IBC 2009 or IBC 201

* Axial live load NOT included in total displayed, or used for overturning resistance, but is included for soil pressure calculation.

Vertical component of active lateral soil pressure IS NOT considered in the calculation of Sliding Resistance.

Vertical component of active lateral soil pressure IS NOT considered in the calculation of Overturning Resistance

Tilt

Horizontal Deflection at Top of Wall due to settlement of soil

(Deflection due to wall bending not considered)

Soil Spring Reaction Modulus	250.0	pci
Horizontal Defl @ Top of Wall (approximate only)	0.120	in

The above calculation is not valid if the heel soil bearing pressure exceeds that of the toe, because the wall would then tend to rotate into the retained soil.

DESIGN EXAMPLE 6	**Page 1 of 4**

TAPERED CONCRETE STEM –

Code: IBC '15

Include seismic effect

Assume lateral support at top of footing

Angle of internal friction

$= \phi = 34°$

Wall friction angle

$= \delta = \dfrac{\varphi}{2} = 17°$

Soil density $= 110$ pcf

$f'_c = 3,000$ psi $\qquad f_y = 60,000$ psi

Backfill slope $= 3:1 = \beta = \quad 18.4°$

Wall friction angle assumed: $\delta = 17°$

Determine seismic factor, k_h

Assume high-seismic California

From charts in IBC for "short period", select $S_s = 1.274$

Then $S_{MS} = F_a S_s$

$F_a = 1.0$ (This is a function of soil characteristics and value of S_s. See IBC '15, Table 1613.3.3 (1).

$\therefore \; S_{MS} = 1.0 \times 1.274 = 1.274$

$S_{DS} = \dfrac{2}{3} S_{MS} = 0.667 \times 1.274 = 0.85$

Per FEMA/NEHRP Part 2, Commentary, 7.5.1.

$k_h = \dfrac{s_{DS}}{2.5} = 0.40 \times 0.85 = 0.34$

Use $k_h = 0.34$ $\qquad\qquad$ $\tan^{-1} \theta = 0.34 = 18.8°$

Determine static and seismic lateral pressures

Use Coulomb/Monokobe-Okabe equations:

$$K_A = \frac{\sin^2 (\phi + 90)}{\sin (90 - \delta) \left[1 + \sqrt{\dfrac{\sin (\phi + \delta) \sin (\phi - \beta)}{\sin (90 - \delta) \sin (\beta + 90)}}\right]^2}$$

$$= \frac{\sin^2 (34 + 90)}{\sin (90 - 17) \left[1 + \sqrt{\dfrac{\sin (34 + 17) \sin (34 - 18.4)}{\sin (90 - 17) \sin (18.4 + 90)}}\right]^2} = 0.328$$

K_A (horizontal) $= \cos \delta \, K_A = 0.96 \times .328 = 0.313$

Tip: Quick solve these – and other – complex equations using the Σ feature on Retain Pro 10.

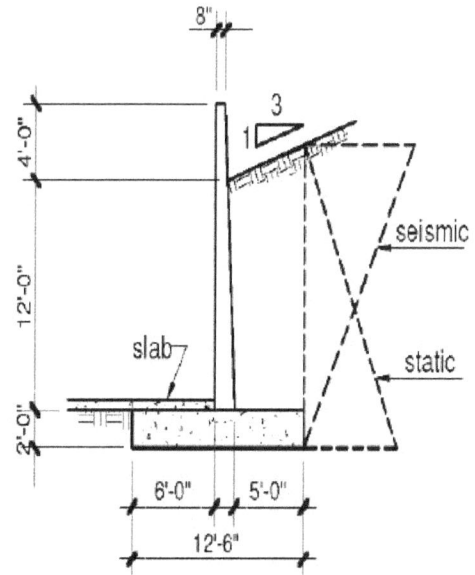

$$K_{AE} = \frac{\sin^2(\phi + 90 - \theta)}{\cos\theta \sin(90 - \theta - \delta)\left[1 + \sqrt{\dfrac{\sin(\phi + \delta)\sin(\phi - \theta - \beta)}{\sin(90 - \delta - \theta)\sin(\beta + 90)}}\right]^2}$$

$$= \frac{\sin^2(34 + 90 - 18.8)}{\cos 18.8 \sin(90 - 18.8 - 17)\left[1 + \sqrt{\dfrac{\sin(34 + 17)\sin(34 - 18.8 - 18.4)}{\sin(90 - 17 - 18.8)\sin(18.4 + 90)}}\right]^2}$$

Because term under radical is zero

$K_{AE} = 1.21$

K_{AE} (horizontal) $= \cos\delta\; K_{AE} = 0.96$ x $1.21 = 1.16$

Static lateral at base of stem $= \dfrac{0.313 \times 110 \times 12^2}{2}$ x $1.6 = 3966^{\#}$ / ft

Seismic portion of lateral $= \dfrac{(1.16 - .313)110 \times 12^2}{2} = 6732^{\#}$ / ft

Static + seismic lateral at base of stem $= 3966 + 6732 = 10,698^{\#}$ / ft

$M_{static\;@\;base} = \dfrac{0.313 \times 110 \times 12^2}{2}$ x $\dfrac{12}{3}$ x $1.6 = 15,865^{'\#}$

$M_{seismic}$ @ stem base assuming point of application $= 0.6H$

   6732 x 0.60 x $12 = 48,470^{'\#}$

Mu @ stem base $= 15,865 + 48,470 = 64,335^{'\#}$

Height to resultant of static and seismic forces

$= \dfrac{15865 + 48,470}{10698} = 6.01$ ft. $\cong$ mid-height

<u>Check Base of Stem</u>

   $M_{static} = 15,865^{'\#}$

   $M_{seismic} = 48,470$ x $1.0 = 48,470^{'\#}$

   Design factored moment $= 15,865 + 48,470 = 64,335^{'\#}$

   Try 18" stem $d = 18.0 - 2.5 = 15.5"$

   A_s required $= 0.17 f_c' d - \sqrt{.029\left(f_c' d\right)^2 - .0063\, f_c'\, M_u}$

      $= .17$ x 3.0 x $15.5 - \sqrt{.029\left(3 \times 15.5\right)^2 - .0063 \times 3 \times 64.335 \times 12}$

      $= 0.965\; \dfrac{sq.\,in.}{ft.}$

DESIGN EXAMPLE 6   **Page 3 of 4**

Check for A_s min. = .0033 x 12 x 15.5 = 0.61

Use #8 @ 9" $A_s = \dfrac{1.0}{0.75} = 1.33 > 0.965$

V_u factored = 10,698$^{\#}$ / ft.

$v = \dfrac{10,698^{\#}}{12\,x\,15.5} = 57.5 < .75\ x\ 2\ x\ \sqrt{3000} = 82.2$ psi

(Note: This example does not include seismic due to stem self-weight. If desired, this can be added as an "additional lateral load," using the appropriate seismic factor.)

Check stem at 4.0 ft. above ftg.   Retained ht. = 8.0 ft.

t = 15.5" (by interpolation)   d = say 13.0"

Use same procedure as base of stem

$M_u = 19,010'\#$

A_s req'd. @ 4' high = $0.17\ x\ 3.0\ x\ 13.0 - \sqrt{.029\ (3\,x\,13)^2 - .0063\ x\ 3\ x\ 19\ x\ 12}$

$= 0.32\ \dfrac{sq.\,in.}{ft.}$

Use #7 @ 18" $A_s = \dfrac{0.60}{1.5} = 0.40 > 0.32$ OK

Stability and footing design:

Total ht. @ back of heel = 12.0 + 2.0 + 5/3 = 15.67'

   (Assume slope starts aligned with stem at bottom).

Static lateral @ bott. of footing = $\dfrac{.313\ x\ 110\ x\ 15.67^2}{2} = 4229\#$ / ft

Seismic lateral @ bott. of footing = $\dfrac{(1.16 - 0.31)\ 110\ x\ 15.67^2}{2}\ x\ 0.71 = 8150^{\#}$

Total lateral = 4229 + 8150 = 12,379$^{\#}$ /ft   (Converted to ASD) ↑

Try 6'-0" toe, 6'-6" heel (incl. stem) = 12'-6" total width

Overturning moment = $\dfrac{(1.16 - .313)\,110\ x\ 15.67^2}{2}\ x\ .6\ x\ 15.67\ x\ 0.71 = 76,310$ ft-lbs.

                              To convert to ASD.   ↑

$+\ \dfrac{.313\ x\ 110\ x\ 15.67^3}{2\ x\ 3} = 76,310 + 22,067 = 98,377$ ft-lbs.

<u>Resisting Moment</u>

(Use vertical comp. of lateral soil force to resist overturning but not to reduce soil pressure).

Resisting moment = (soil over heel) (arm) + (sloped soil over heel) (arm) + (stem wt.) (arm) + (earth @ stem) (arm) + (ftg. wt.) (arm) + (vert. comp. @ back of heel) (arm)

= (5 x 110 x 12) (10) + (5.63 x 1.88 x .5 x 110) (10.63) + (16 x 1.08 x 150) x (6.57)

+ (12 x .63 x .5 x 110) (7.29) + (12.5 x 2.0 x 150) (6.25)

+ 4229 tan 17°) (12.5) = 131,543 ft-lbs. (Vert. comp. = 4229 tan 17° = 1293$^{\#}$

Total vert. load = (5 x 110 x 12) + (5.63 x 1.88 x .5 x 110) + (16 x 1.08 x 150)

+ (12 x .83 x .5 x 110) + (12.5 x 2 x 150) + 1293$^{\#}$ = 15,365$^{\#}$

Overturning ratio = $\dfrac{131,543}{98,377}$ = 1.34

Soil pressure: (Vert. Component not used)

$\bar{x} = \dfrac{131,543 - 98,377 - (1293\,x\,12.5)}{15,365 - 1293}$ = 1.21 ft.

Eccentricity = $(12.5/2)$ - 1.21 = 5.04' = 60.5"

e for inside middle third = $12.5/6$ = 2.08, 5.04 < 2.08 ∴ Outside middle third

Soil pressure = $\dfrac{15,365 - 1293}{.75\,x\,12.5 - 1.5\,x\,5.04}$ = 7753 psf

Sliding S.F. = $\dfrac{0.4\,x\,(15,365 - 1293)}{12,379}$ = 0.45

However, not applicable since slab is present and to be designed to resist lateral.

DESIGN EXAMPLE 6 Report Printout

Basics of Retaining Wall Design   Title   EX-6 :   Dsgn: HB   Date   2 JUL 2018   Page   1
Job #
Description
EX-6

This Wall In File: c:\Users\chris\Dropbox\RetainPro 10 Project Files\HB\Hugh.RPX

RetainPro (c) 1987-2018, Build 11.18.08.22
License : KW-06000216
License To : RetainPro Development Center

Tapered Stem Concrete Retaining Wall   Code: IBC 2015,ACI 318-14,ACI 530-13

Criteria

Retained Height	=	12.00 ft
Wall height above soil	=	4.00 ft
Slope Behind Wall	=	3.00
Height of Soil over Toe	=	0.00 in
		ft

Soil Data

Allow Soil Bearing	=	8,000.0 psf
Coulomb Soil Pressure calculation		
Soil Friction Angle	=	34.0 deg
Active Pressure Ka*Gamma	=	34.2 psf/ft
Passive Pressure:Kp*Gamma		711.9 psf/ft
Soil Density	=	110.00 pcf
Footing||Soil Friction	=	0.400
Soil height to ignore for passive pressure	=	0.00 in

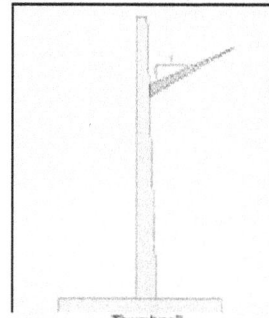

Thumbnail

Surcharge Loads

Surcharge Over Heel	=	0.0 psf
>>>Used To Resist Sliding & Overturning		
Surcharge Over Toe	=	0.0 psf
Used for Sliding & Overturning		

Axial Load Applied to Stem

Axial Dead Load	=	0.0 lbs
Axial Live Load	=	0.0 lbs
Axial Load Eccentricity	=	0.0 in

Lateral Load Applied to Stem

Lateral Load	=	0.0 #/ft
Height to Top	=	0.00 ft
Height to Bottom	=	0.00 ft
Wind on Exposed Stem	=	0.0 psf

Adjacent Footing Load

Adjacent Footing Load	=	0.0 lbs
Footing Width	=	0.00 ft
Eccentricity	=	0.00 in
Wall to Ftg CL Dist	=	0.00 ft
Footing Type		Line Load
Base Above/Below Soil at Back of Wall	=	0.0 ft
Poisson's Ratio	=	0.300

Earth Pressure Seismic Load

Design Kh	=	0.340 g

Using Mononobe-Okabe / Seed-Whitman procedure

Kae for seismic earth pressure	=	1.333
Ka for static earth pressure	=	0.314
Difference Kae - Ka	=	1.019

Added seismic base force   13756.3 lbs

<<—— Note! These are horizontal components

Stem Weight Seismic Load

Fp/Wp Weight Multiplier	=	0.000 g

Added seismic base force   0.0 lbs

Design Summary

Total Bearing Load	=	16,772 lbs
resultant ecc.	=	75.97 in
Resultant Exceeds Ftg Width!		
Soil Pressure @ Toe	=	0 psf OK
Soil Pressure @ Heel	=	0 psf OK
Allowable	=	8,000 psf
ACI Factored @ Toe	=	0 psf
ACI Factored @ Heel	=	0 psf
Footing Shear @ Toe	=	5.9 psi OK
Footing Shear @ Heel	=	57.4 psi OK
Allowable	=	82.2 psi
Sliding Stability Ratio	=	0.45 UNSTABL

Sliding Calcs

Lateral Sliding Force	=	18,100.5 lbs
less 100% Passive Force	-	1,423.9 lbs
less 100% Friction Force	-	6,708.6 lbs
Added Force Req'd	=	9,968.1 lbs NG
for 1.5 Stability	=	19,016.3 lbs NG

Vertical component of active lateral soil pressure IS considered in the calculation of soil bearing pressures

Load Factors

Building Code	IBC 2015,ACI
Dead Load	1.200
Live Load	1.600
Earth, H	1.600
Wind, W	1.000
Seismic, E	1.000

Tapered Concrete Stem Design Data

Thickness at TOP	=	8.00 in	Fy =	60,000 psi
Thickness at BOTTOM	=	18.00 in	fc =	3,000 psi

Rebar Cover (rebar center to concrete face)2.00 in

		@ Height #1	@ Height #2	@ Base of Wall
		Bending NG!	Bending NG!	Bending NG!
Design Height Above Ftg	=	8.00 ft	4.00 ft	0.00 ft
Rebar Size	=	# 5	# 9	# 9
Rebar Spacing	=	12.00 in	12.00 in	6.00 in
Rebar Depth \d	=	10.50 in	13.00 in	15.50 in
Design Data				
Mu ...Actual	=	22,144.2 ft-#	80,560.7 ft-#	163225.0 ft-#
Mn * Phi ...Allowable	=	14,222.3 ft-#	54,075.0 ft-#	121600.0 ft-#
Shear Force @ this height	=	9,421.2 lbs	17,561.2 lbs	24,419.9 lbs
Vu ...Actual	=	74.77 psi	112.57 psi	131.29 psi
Vn * Phi ...Allowable	=	82.16 psi	82.16 psi	82.16 psi

DESIGN EXAMPLE 6 — Report Printout

Tapered Stem Concrete Retaining Wall — Code: IBC 2015, ACI 318-14, ACI 530-13

Footing Strengths & Dimensions

Toe Width = 6.00 ft
Heel Width = 6.50
Total Footing Width = 12.50
Footing Thickness = 24.00 in
Key Width = 0.00 in
Key Depth = 0.00 in
Key Distance from Toe = 4.00 ft
f'c = 3,000 psi   Fy = 60,000 psi
Footing Concrete Density = 150.00 pcf
Min. As % = 0.0018
Cover @ Top = 2.00 in   @ Btm = 3.00 in

Footing Design Results

	Toe	Heel
Factored Pressure	0	0 psf
Mu: Upward	0	0 ft-#
Mu: Downward	6,480	48,765 ft-#
Mu: Design	-6,480	48,765 ft-#
Actual 1-Way Shear	5.85	57.35 psi
Allow 1-Way Shear	43.82	82.16 psi

Toe Reinforcing = #4 @ 18.00 in
Heel Reinforcing = #6 @ 12.00 in
Key Reinforcing = #5 @ 14.35 in

Other Acceptable Sizes & Spacings
Toe: Not req'd, Mu < phi*5*lambda*sqrt(f'c)*Sm
Heel: #4@ 3.52 in, #5@ 5.45 in, #6@ 7.74 in, #7@ 10.55 in, #8@ 13.90 in, #9@ 17
Key: No key defined

Summary of Overturning & Resisting Forces & Moments

Item	OVERTURNING Force lbs	Distance ft	Moment ft-#		RESISTING Force lbs	Distance ft	Moment ft-#
Heel Active Pressure	4,344.1	5.31	23,088.2	Soil Over Heel	6,600.0	10.00	66,000.0
Surcharge Over Toe				Sloped Soil Over Heel	580.1	10.63	6,163.3
Adjacent Footing Load				Surcharge Over Heel			
Added Lateral Load				Adjacent Footing Load			
Load @ Stem Above Soil				Axial Dead Load on Stem		0.00	
Seismic Load	13,756.3	9.40	129,309.7	Soil Over Toe			
				Surcharge Over Toe			
Total	18,100.5	O.T.M. =	152,397.8	Stem Weight	2,600.0	6.57	17,077.8
				Earth above Sloping Stem	412.5	7.29	3,007.8
Resisting/Overturning Ratio		=	0.99	Footing Weight	3,750.0	6.25	23,437.5
				Key Weight		4.00	
Vertical Loads used for Soil Pressure =			16,771.5 lbs	Vert. Component	2,828.9	12.50	35,361.7
				Total = 16,771.5 lbs	R.M. =		151,048.2

Vertical component of active lateral soil pressure IS considered in the calculation of Sliding Resistance.

Vertical component of active lateral soil pressure IS considered in the calculation of Overturning Resistance.

Tilt

Horizontal Deflection at Top of Wall due to settlement of soil

(Deflection due to wall bending not considered)

Soil Spring Reaction Modulus = 250.0 pci
Horizontal Defl @ Top of Wall (approximate only) = 0.120 in

The above calculation is not valid if the heel soil bearing pressure exceeds that of the toe, because the wall would then tend to rotate into the retained soil.

DESIGN EXAMPLE 7   Page 1 of 2

RESTRAINED CMU WALL
Design Data

Code: IBC '15

Soil bearing = 2000 psf

Soil density = 110 pcf

EFP = 30 psf/ft.

Passive - (not appl. because of floor slab)

f'_m = 1500 psi

f_s = 24,000 psi

f_y = 60,000 psi

f'_c = 2500 psi

w   = wind = 15 psf

P   = 144 plf DL

e   = 7.0 in.

Assume 100% fixity at base

Assume lateral restraint at top of footing to resist sliding

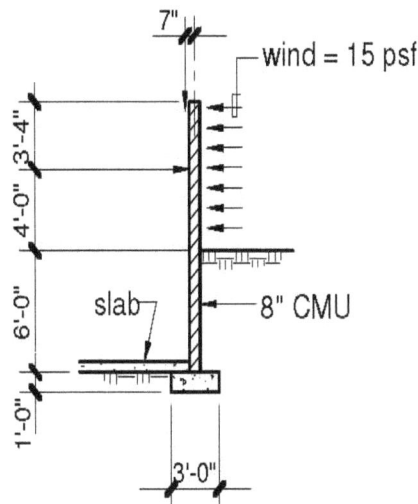

*Moments, Shear, and Reaction

R   =   reaction @ top restraint = 160 lbs.   V at base = 685 lbs.

M @ top restraint = 144 (7 / 12) + 15 x $(3.33)^2$ / 2 = 167 [']#

M @ base = 7488 in. lbs. = 624[#]

+M Max. = 2592 in. lbs. = 216[#]   @ h = 5.84 ft.

Check stem @ base

M = 7488 in. lbs. = 624 ft. lbs.

* (Obtained from Single Span Beam Analysis program in Enercalc's Structural Engineering Library, Version 6.0.)

Try 8" CMU, #5 @ 32", d = 5.3", solid grouted, n = 21.5

f'_m = 1500 psi f_s = 24,000 f_b = .33 x 1500 = 500 psi

Use #5 @ 32" o.c. @ edge $np = \dfrac{21.5 \; x \; .31}{32 \; x \; 5.25} = 0.040$

$\dfrac{2}{kj} = 8.9$ j = 0.92

$M_m = \dfrac{500 \; x \; 12 \; x \; 5.25^2}{8.9} \; x \dfrac{1}{12} = 1548$ ft. lbs. > 624   OK

$$M_s = 24,000 \times \frac{.31}{2.67} \times .92 \times 5.25 \times \frac{1}{12} = 1122^{'\#} > 624$$

$$V = 490 \text{ lbs.} \qquad v = \frac{490}{12 \times 5.25} = 7.78 \qquad < v_{allow} = 38.7$$

Check @ Max. Positive Moment

+M = 2592 in. lbs. = 216 ft. lbs.

Use #5 @ 32" o.c. @ center

$$np = \frac{21.5 \times 0.31}{32 \times 3.75} = 0.056 \qquad \frac{2}{kj} = 7.80 \qquad j = 0.91$$

$$M_m = \frac{500 \times 12 \times 3.75^2}{7.80} \times \frac{1}{12} = 901 \text{ ft. lbs.} > 216 \quad \text{OK}$$

$$M_s = 24,000 \times \frac{0.31}{2.67} \times 0.91 \times 3.75 \times \frac{1}{12} = 792 \text{ ft. lbs.} > 216$$

Check Moment @ Lateral Support

M = 167 '#

Use #5 @ 32" o.c. @ center

OK per above analysis for positive mid-height moment.

Soil Bearing

$$\text{Embedment of hooked bar in footing} = \frac{0.02 \times 60,000 \times .625 \times 0.7}{\sqrt{2000}} \times \frac{624}{1122} = 6.5"$$

Min. ftg. thickness = 6.6 + 3.0 = 9.6 in.

Try 3'-0" wide ftg. Centered under stem   Use 12" thick > 9.6"

Total vertical load = 13.33 x 78 psf + 6 x 1.17 x 110 + 3.0 x 1.0 x 150 + 144 (DL only)

= 2406$^{\#}$ M @ base = 624$^{'\#}$

Distance from toe to centroid of soil pressure

$$= \frac{13.33' \times 78 \times 1.5 + 1.17 \times 6 \times 110 \times 2.42 + 3' \times 1' \times 150 \times 1.5 + 144 \times .92 - 624}{2406}$$

= 1.41 ft. $\qquad e = \frac{3.0}{2} - 1.41 = 0.09 \text{ ft. (in middle third)}$

$$\text{Soil pressure} = \frac{2406}{3.0} \pm \frac{2406 \times .00 \times 6}{3^2} = 802 \pm 0 = 802 \text{ psf max. (uniform)}$$

DESIGN EXAMPLE 7	Report Printout

Basics of Retaining Wall Design

Title: EX-7
Job #:
Description ....
EX-7

Degnr: HB   Date: 2 JUL 2018   Page: 1

This Wall in File: c:\Users\chris\Dropbox\RetainPro 10 Project Files\HB\Hugh.RPX

RetainPro (c) 1987-2018, Build 11.18.06.22
License : KW-06060216
License To : RetainPro Development Center

Restrained Retaining Wall   Code: IBC 2015,ACI 318-14,ACI 530-13

Criteria

Retained Height	= 6.00 ft
Wall height above soil	= 7.33 ft
Total Wall Height	= 13.33 ft
Top Support Height	= 10.00 ft
Slope Behind Wall	= 0.00
Height of Soil over Toe	= 0.00 in

Soil Data

Allow Soil Bearing	= 2,000.0 psf
Equivalent Fluid Pressure Method	
At-rest Heel Pressure	= 30.0 psf/ft
Passive Pressure	= 250.0 psf/ft
Soil Density	= 110.00 pcf
Footing||Soil Friction	= 0.400
Soil height to ignore for passive pressure	= 0.00 in

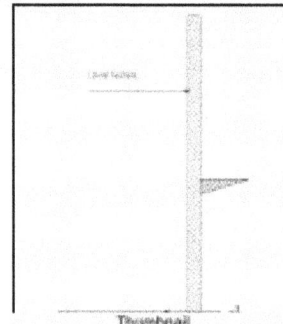

Thumbnail

Surcharge Loads

Surcharge Over Heel	= 0.0 psf
>>>Used To Resist Sliding & Overturning	
Surcharge Over Toe	= 0.0 psf
Used for Sliding & Overturning	

Axial Load Applied to Stem

Axial Dead Load	= 144.0 lbs
Axial Live Load	= 0.0 lbs
Axial Load Eccentricity	= 7.0 in

Earth Pressure Seismic Load

K_h Soil Density Multiplier = 0.200 g   Added seismic per unit area = 0.0 psf

Stem Weight Seismic Load

F_p / W_p Weight Multiplier = 0.000 g   Added seismic per unit area = 0.0 psf

Uniform Lateral Load Applied to Stem

Lateral Load	= 0.0 #/ft
.. Height to Top	= 0.00 ft
.. Height to Bottom	= 0.00 ft
Load Type	= Wind (W)
Wind on Exposed Stem	= 15.0 psf

Adjacent Footing Load

Adjacent Footing Load	= 0.0 lbs
Footing Width	= 0.00 ft
Eccentricity	= 0.00 in
Wall to Ftg CL Dist	= 0.00 ft
Footing Type	Line Load
Base Above/Below Soil at Back of Wall	= 0.0 ft
Poisson's Ratio	= 0.300

Design Summary

Total Bearing Load	= 2,402 lbs
... resultant ecc.	= 0.00 in
Soil Pressure @ Toe	= 801 psf OK
Soil Pressure @ Heel	= 801 psf OK
Allowable	= 2,000 psf
Soil Pressure Less Than Allowable	
ACI Factored @ Toe	= 961 psf
ACI Factored @ Heel	= 961 psf
Footing Shear @ Toe	= 9.0 psi OK
Footing Shear @ Heel	= 0.1 psi OK
Allowable	= 67.1 psi
Reaction at Top	= 158.7 lbs
Reaction at Bottom	= 684.7 lbs

Sliding Calcs
Lateral Sliding Force = 684.7 lbs

Vertical component of active lateral soil pressure IS NOT considered in the calculation of soil bearing

Load Factors

Building Code	IBC 2015,ACI
Dead Load	1.200
Live Load	1.600
Earth, H	1.600
Wind, W	1.000
Seismic, E	1.000

Masonry Stem Construction

Thickness	= 8.00 in	f'm	= 1,500 psi	Short Term Factor	=	1.000
Wall Weight	= 78.0 psf	Fs	= 32,000 psi	Equiv. Solid Thick	=	7.600 in
Stem is FIXED to top of footing				n Ratio (Es/Em)	=	21.481
Block Type = Medium Weight						
Design Method = ASD						
Solid Grouted						

	@ Top Support	Mmax Between Top & Base	@ Base of Wall
	As < Min %	Stem OK	As < Min %
Design Height Above Ftg =	10.00 ft	4.22 ft	0.00 ft
Rebar Size	# 5	# 5	# 5
Rebar Spacing	32.00 in	32.00 in	32.00 in
Rebar Placed at	Center	Center	Edge
Rebar Depth 'd'	3.75 in	3.75 in	5.25 in
Design Data			
fb/FB + fa/Fa	1.000	0.205	1.000
Moment....Actual	167.2 ft-#	215.6 ft-#	623.9 ft-#
Moment....Allowable	1,053.1 ft-#	1,053.1 ft-#	1,404.8 ft-#
Shear Force @ this height	110.3 lbs		480.7 lbs
Anet	91.50 in2		91.50 in2
Shear....Actual	1.21 psi		5.35 psi
Shear....Allowable	43.82 psi		43.82 psi

Other Acceptable Sizes & Spacings:
Toe: # 7 @ 18.00 in   -or- Not req'd: Mu < phi*5*lambda*sqrt(f'c)*Sm
Heel: # 6 @ 18.00 in   -or- Not req'd: Mu < phi*5*lambda*sqrt(f'c)*Sm
Key: No key defined   -or- No key defined

DESIGN EXAMPLE 7   Report Printout

Basics of Retaining Wall Design

Title   EX-7
Job #
Description
EX-7

Dsgnr   HB

Date   2 JUL 2018

Page   2

This Wall in File: c:\Users\chris\Dropbox\RetainPro 10 Project Files\HB\Hugh.RPX

RetainPro (c) 1987-2018, Build 11.18.06.22
License : KW-06000215
License To : RetainPro Development Center

Restrained Retaining Wall   Code: IBC 2015,ACI 318-14,ACI 530-13

Footing Strengths & Dimensions

Toe Width	=	1.17 ft
Heel Width	=	1.83
Total Footing Width	=	3.00
Footing Thickness	=	12.00 in
Key Width	=	0.00 in
Key Depth	=	0.00 in
Key Distance from Toe	=	0.00 ft

f'c = 2,000 psi   Fy = 60,000 psi
Footing Concrete Density = 150.00 pcf
Min As % = 0.0018
Cover @ Top = 2.00 in   @ Btm = 3.00 in

Footing Design Results

	Toe	Heel
Factored Pressure	=	961	961 psf
Mu' Upward	=	0	850 ft-#
Mu' Downward	=	0	860 ft-#
Mu Design	=	256	10 ft-#
Actual 1-Way Shear	=	8.95	0.12 psi
Allow 1-Way Shear	=	67.08	67.08 psi

Min footing T&S reinf Area	0.78	in2
Min footing T&S reinf Area per foot	0.26	in2 .ft

If one layer of horizontal bars   If two layers of horizontal bars
#4@ 9.26 in   #4@ 18.52 in
#5@ 14.35 in   #5@ 28.70 in
#6@ 20.37 in   #6@ 40.74 in

Summary of Forces on Footing : Slab RESISTS sliding, stem is FIXED at footing

Forces acting on footing for soil pressure   >>> Sliding Forces are restrained by the adjacent slab
Load & Moment Summary For Footing : For Soil Pressure Calcs

Moment @ Top of Footing Applied from Stem	=			-623.9 ft-#
Surcharge Over Heel	=	lbs	ft	ft-#
Adjacent Footing Load	=	lbs	ft	ft-#
Axial Dead Load on Stem	=	144.0 lbs	1.50 ft	216.5 ft-#
Soil Over Toe	=	lbs	ft	ft-#
Surcharge Over Toe	=	lbs	ft	ft-#
Stem Weight	=	1,039.7 lbs	1.50 ft	1,563.1 ft-#
Soil Over Heel	=	767.8 lbs	2.42 ft	1,856.8 ft-#
Footing Weight	=	450.0 lbs	1.50 ft	675.0 ft-#
Total Vertical Force	=	2,401.5 lbs	Base Moment =	3,687.4 ft-#

Soil Pressure Resulting Moment =   0.0 ft-#

Vertical component of active lateral soil pressure IS NOT considered in
the calculation of Sliding Resistance.

DESIGN EXAMPLE 8   Page 1 of 2

RESTRAINED CONCRETE WALL

Code: IBC '15

Tie-back @ 16 ft. high

Use EFP = 40 pcf

Backfill slope = 3:1

Soil bearing = 3,000 psf

Soil Wt = 110 pcf

$f_c' = 3,000$ psi $f_y = 60,000$ psi

Slab lateral restraint @ base

Assume stem "pinned" at footing

Reactions:

$$W = \frac{40 \times 20^2}{2} = 8,000 \text{ lbs.}$$

$$R @ \text{tie-back} = \frac{8000 \times \left(\frac{20}{3}\right)}{16}$$

$$= 3333 \; \frac{lbs.}{ft.}$$

R @ base = [40 x (20 +1)^2 / 2] - 3333

     = 5487 lbs.

Moments:

Dist. to max mom where V = 0: = 7.09′ by statics

Max. pos. moment = 15,250$^{\#}$ x 1.6 = M_u = 24,400$^{\#}$

Design @ max. positive moment

M_u = 24,400$^{"\#}$ Try 12"   (d = 10")

A_s required = $.17 \times 3 \times 10 - \sqrt{.029\,(3 \times 10\,)^2 \; - \; .0063 \times 3 \times 24.4 \times 12}$

= 0.57 sq. in./ft.

Use #7 @ 9" o.c. $A_s = \dfrac{.60}{.75} = 0.80 > 0.57$

Design moment at support = 426$^{\#}$ x 1.6 = 681$^{\#}$

Use min. vert. reinf. @ center throughout   O.K. by inspection

Min. vert. reinf. = $\dfrac{200}{f_y}$ = .0033 sq. in.

#6 @ 18" o.c. @ center vert. $A_s = \dfrac{.44}{(6 \times 18)}$ = .0041 > .0033

Total vert. load to ftg.

     = 20 x 150 + .50 x 20 x 110 = 4100 lbs.

Try 2′-0" wide ftg. Assume "pin" connection wall to footing.

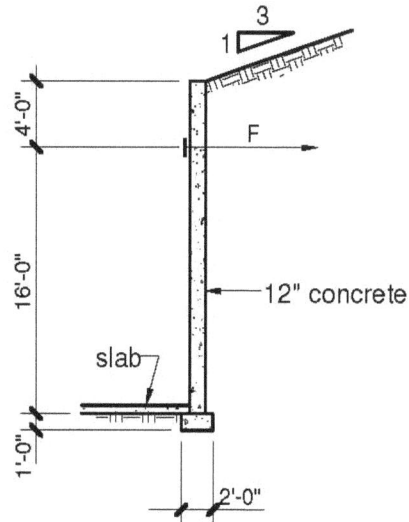

DESIGN EXAMPLE 8   **Page 2 of 2**

Moments about front edge of ftg.

$$= (20 \times 150)\,1.0 + (.50 \times 20 \times 110)\,1.75 + (2 \times 150 \times 1.0)$$

$$= 5225^{'\#}$$

Ecc. $= \left[(5225/4400) - (2/2) \right] \times 12 = 2.25"$

Mom. @ stem-ftg. interface due to ecc. $= 4400 \times \dfrac{2.25}{12} = 825$ ft. lbs.

Allow. Mom. @ stems – ftg. interface:

$$a \quad = \frac{.31 \times 60{,}000}{1.5 \times 0.85 \times 3000 \times 12} \quad = 0.41$$

$$\phi M_n \quad = 0.9 \left[\frac{.31}{1.5} \times 60{,}000 \left(6 - \frac{0.40}{2} \right) \times \frac{1}{12} \right] = 5394 \text{ ft. lbs.}$$

Since base of stem allow. moment exceeds mom. due to ftg. eccentricity, soil pressure is uniform $= \dfrac{4400}{2} = 2200$ psf.

Note: If stem ftg. mom. < 825 ft. lbs., then:

$$\text{Soil pressure} \quad = \frac{4400}{2} \pm \frac{4400 \times \left(2.25/12 \right) \times 6}{2^2}$$

$$= 2200 \pm 1238$$

$$= 962 \text{ psf @ toe}$$

$$= 3438 \text{ psf @ heel}$$

Note: Check slab for this lateral force of 5487 lbs. – usually resisted by sliding friction.

DESIGN EXAMPLE 8

Report Printout

Basics of Retaining Wall Design

Title EX-8
Job #
Description ...
EX-8

Dsgnr: HB

Date: 2 JUL 2018

Page: 1

This Wall in File: c:\Users\chris\Dropbox\RetainPro 10 Project Files\HB\Hugh.RPX

RetainPro (c) 1987-2018, Build 11.18.06.22
License : KW-06000218
License To : RetainPro Development Center

Restrained Retaining Wall

Code: IBC 2015, ACI 318-14, ACI 530-13

Criteria

Retained Height	=	20.00 ft
Wall height above soil	=	0.00 ft
Total Wall Height	=	20.00 ft
Top Support Height	=	16.00 ft
Slope Behind Wall	=	3.00
Height of Soil over Toe	=	0.00 in

Soil Data

Allow Soil Bearing	=	3,000.0 psf
Equivalent Fluid Pressure Method		
At-rest Heel Pressure	=	40.0 psf/ft
	=	
Passive Pressure	=	380.0 psf/ft
Soil Density	=	110.00 pcf
Footing||Soil Friction	=	0.400
Soil height to ignore for passive pressure	=	12.00 in

Thumbnail

Surcharge Loads

Surcharge Over Heel	=	0.0 psf
>>>Used To Resist Sliding & Overturning		
Surcharge Over Toe	=	0.0 psf
Used for Sliding & Overturning		

Axial Load Applied to Stem

Axial Dead Load	=	0.0 lbs
Axial Live Load	=	0.0 lbs
Axial Load Eccentricity	=	0.0 in

Earth Pressure Seismic Load

Stem Weight Seismic Load

Uniform Lateral Load Applied to Stem

Lateral Load	=	0.0 #/ft
Height to Top	=	0.00 ft
Height to Bottom	=	0.00 ft
Load Type	=	Wind (W)
		(Strength Level)
Wind on Exposed Stem	=	0.0 psf

K_h Soil Density Multiplier = 0.200 g

F_p / W_p Weight Multiplier = 0.000 g

Adjacent Footing Load

Adjacent Footing Load	=	0.0 lbs
Footing Width	=	0.00 ft
Eccentricity	=	0.00 in
Wall to Ftg CL Dist	=	0.00 ft
Footing Type		Line Load
Base Above/Below Soil at Back of Wall	=	0.0 ft
Poisson's Ratio	=	0.300

Added seismic per unit area = 0.0 psf

Added seismic per unit area = 0.0 psf

Design Summary

Total Bearing Load	=	4,405 lbs
...resultant ecc.	=	2.24 in
Soil Pressure @ Toe	=	972 psf OK
Soil Pressure @ Heel	=	3,433 psf NG
Allowable	=	3,000 psf
Soil Pressure Exceeds Allowable!		
ACI Factored @ Toe	=	1,166 psf
ACI Factored @ Heel	=	4,120 psf
Footing Shear @ Toe	=	0.4 psi OK
Footing Shear @ Heel	=	4.0 psi OK
Allowable	=	82.2 psi
Reaction at Top	=	3,322.0 lbs
Reaction at Bottom	=	5,486.7 lbs

Sliding Calcs

Lateral Sliding Force	=	5,486.7 lbs

Vertical component of active lateral soil pressure IS NOT considered in the calculation of soil bearing.

Load Factors

Building Code	IBC 2015, ACI
Dead Load	1.200
Live Load	1.600
Earth, H	1.600
Wind, W	1.000
Seismic, E	1.000

Concrete Stem Construction

Thickness	=	12.00 in
Wall Weight	=	150.0 psf

Fy = 60,000 psi
f'c = 3,000 psi

Stem is FREE to rotate at top of footing

		@ Top Support	Mmax Between Top & Base	@ Base of Wall
			As = Min %	
Design Height Above Ftg	=	16.00 ft	7.13 ft	0.00 ft
Rebar Size		# 5	# 7	# 5
Rebar Spacing		18.00 in	9.00 in	18.00 in
Rebar Placed at		Center	Edge	Center
Rebar Depth \d		6.00 in	10.00 in	6.00 in
Design Data				
fb/FB + fa/Fa	=	1.000	0.741	1.000
Mu...Actual	=	662.7 ft-#	24,568.7 ft-#	0.0 ft-#
Mn * Phi....Allowable	=	5,391.0 ft-#	33,168.0 ft-#	5,391.0 ft-#
Shear Force @ this height	=	4,821.3 lbs		7,466.7 lbs
Shear....Actual	=	66.96 psi		103.70 psi
Shear... Allowable	=	82.16 psi		82.16 psi

Other Acceptable Sizes & Spacings:

Toe: None Spec'd	-or-	Not req'd: Mu < phi*5*lambda*sqrt(f'c)*Sm
Heel: None Spec'd	-or-	Not req'd: Mu < phi*5*lambda*sqrt(f'c)*Sm
Key: No key defined	-or-	No key defined

DESIGN EXAMPLE 8 Report Printout

Basics of Retaining Wall Design

Title   EX-8
Job #
Description ...
EX-8

Dsgnr   HB

Date   2 JUL 2018

Page   2

This Wall in File: c:\Users\chris\Dropbox\RetainPro 10 Project Files\HB\Hugh.RPX

RetainPro (c) 1987-2018, Build 11.18.08.22
License : KW-06006216
License To : RetainPro Development Center

Restrained Retaining Wall

Code : IBC 2015,ACI 318-14,ACI 530-13

Concrete Stem Rebar Area Details

Top Support

	Vertical Reinforcing	Horizontal Reinforcing	
As (based on applied moment) :	0.0267 in2/ft		
(4/3) * As	0.0356 in2/ft	Min Stem T&S Reinf Area 4.608 in2	
200bd/fy : 200(12)(6)/60000	0.24 in2/ft	Min Stem T&S Reinf Area per ft of stem Height 0.288 in2/ft	
0.0018bh : 0.0018(12)(12)	0.2592 in2/ft	Horizontal Reinforcing Options	
------------		One layer of :	Two layers of :
Required Area :	0.2592 in2/ft	#4@ 8.33 in	#4@ 16.67 in
Provided Area :	0.2067 in2/ft	#5@ 12.92 in	#5@ 25.83 in
Maximum Area :	0.9754 in2/ft	#6@ 18.33 in	#6@ 36.67 in

Mmax Between Ends

	Vertical Reinforcing	Horizontal Reinforcing	
As (based on applied moment) :	0.5639 in2/ft		
(4/3) * As	0.7519 in2/ft	Min Stem T&S Reinf Area 2.554 in2	
200bd/fy : 200(12)(10)/60000	0.4 in2/ft	Min Stem T&S Reinf Area per ft of stem Height 0.288 in2/ft	
0.0018bh : 0.0018(12)(12)	0.2592 in2/ft	Horizontal Reinforcing Options	
------------		One layer of :	Two layers of :
Required Area :	0.5639 in2/ft	#4@ 8.33 in	#4@ 16.67 in
Provided Area :	0.8 in2/ft	#5@ 12.92 in	#5@ 25.83 in
Maximum Area :	1.6256 in2/ft	#6@ 18.33 in	#6@ 36.67 in

Base Support

	Vertical Reinforcing	Horizontal Reinforcing	
As (based on applied moment) :	0 in2/ft		
(4/3) * As	0 in2/ft	Min Stem T&S Reinf Area 2.054 in2	
200bd/fy : 200(12)(6)/60000	0.24 in2/ft	Min Stem T&S Reinf Area per ft of stem Height 0.288 in2/ft	
0.0018bh : 0.0018(12)(12)	0.2592 in2/ft	Horizontal Reinforcing Options	
------------		One layer of :	Two layers of :
Required Area :	0.2592 in2/ft	#4@ 8.33 in	#4@ 16.67 in
Provided Area :	0.2067 in2/ft	#5@ 12.92 in	#5@ 25.83 in
Maximum Area :	0.9754 in2/ft	#6@ 18.33 in	#6@ 36.67 in

Footing Strengths & Dimensions

Toe Width	=	0.50 ft
Heel Width	=	1.50
Total Footing Width	=	2.00
Footing Thickness	=	12.00 in
Key Width	=	0.00 in
Key Depth	=	0.00 in
Key Distance from Toe	=	0.00 ft

f'c = 3,000 psi   Fy = 60,000 psi
Footing Concrete Density = 150.00 pcf
Min. As % = 0.0018
Cover @ Top = 2.00 in   @ Btm = 3.00 in

Footing Design Results

		Toe	Heel
Factored Pressure	=	1,166	4,120 psf
Mu' Upward	=	177	484 ft-#
Mu' Downward	=	23	354 ft-#
Mu : Design	=	154	-130 ft-#
Actual 1-Way Shear	=	0.37	4.03 psi
Allow 1-Way Shear	=	82.16	82.16 psi

Min footing T&S reinf Area   0.52   in2
Min footing T&S reinf Area per foot   0.26   in2 /ft
If one layer of horizontal bars: If two layers of horizontal bars:
#4@ 9.26 in   #4@ 18.52 in
#5@ 14.35 in   #5@ 28.70 in
#6@ 20.37 in   #6@ 40.74 in

DESIGN EXAMPLE 8   Report Printout

Basics of Retaining Wall Design

Title: EX-8	Dsgnr: HB	Page: 3
Job #	Date: 2 JUL 2018	
Description ...		
EX-8		

This Wall in File: c:\Users\chris\Dropbox\RetainPro 10 Project Files\HB\Hugh.RPX

RetainPro (c) 1987-2018, Build 11.18.06.22
License : KW-06060215
License To : RetainPro Development Center

Restrained Retaining Wall

Code: IBC 2015,ACI 318-14,ACI 530-13

Summary of Forces on Footing : Slab RESISTS sliding, stem is PINNED at footing

Forces acting on footing soil pressure

(taking moments about front of footing to find eccentricity)

Surcharge Over Heel	=	lbs	ft	ft-#
Axial Dead Load on Stem	=	lbs	0.00 ft	ft-#
Soil Over Toe	=	lbs	ft	ft-#
Adjacent Footing Load	=	lbs	ft	ft-#
Surcharge Over Toe	=	lbs	ft	ft-#
Stem Weight	=	3,000.0 lbs	1.00 ft	3,000.0 ft-#
Soil Over Heel	=	1,100.0 lbs	1.75 ft	1,925.0 ft-#
Footing Weight	=	300.0 lbs	1.00 ft	300.0 ft-#
Total Vertical Force	=	4,404.6 lbs	Moment =	5,225.0 ft-#

Net Mom. at Stem/Ftg Interface =	-820.4 ft-#
Allow. Mom. @ Stem/Ftg Interface =	3,369.4 ft-#
Allow. Mom. Exceeds Applied Mom.?	Yes
Therefore Uniform Soil Pressure =	2,202.3 psf

Vertical component of active lateral soil pressure IS NOT considered in the calculation of Sliding Resistance

DESIGN EXAMPLE 9   Page 1 of 3

<u>Gravity Wall</u>:

Retained height = 6.0 ft.
Wt. of rubble masonry = 145 pcf
Allow. comp. = 100 psi no tension > 10 psi
Allow. passive = 300 pcf
Soil bearing = 2000 psf
Cohesion = 200 psf
Width of stem base = 30 in.

$$S @ \text{ base} = \frac{12 \times 30^2}{6} = 1800 \; in.^3$$

Backfill slope = 2:1
Soil density = 110 pcf
EFP = 43 pcf

$$M_{base} = \frac{43 \times 6^3}{2 \times 3} = 1548 \; ft - lbs.$$

Wt. of stem above base (rubble + earth over back face)

     = 145 (1 x 6 + 1 x 6 x 0.5 + 0.5 x 6 x 0.5)
     = 1523 lbs.

$$\underline{\text{Stress @ base:}} \; = \frac{1523}{12 \times 30} \pm \frac{1548 \times 12}{1800}$$

     = 4.23 ± 10.32
Max. comp = 14.55 psi
Max. tension = 6.09 psi OK

$$v @ \text{ base} = \frac{43 \times 6^2 \times 0.5}{12 \times 30} = 2.15 \; psi \;\; OK$$

<u>Check @ 2'-0" height above base</u>
Thickness = 24"   S = 1152 in^2

$$M @ 2' \text{ high} = \frac{43 \times 4^3}{2 \times 3} = 459 \; ft - lbs.$$

Wt. above 2' = 145 (1 x 4 + .67 x 4 x 0.5 +
     .33 x 4 x 0.5) = 870 lbs.

$$\text{Stress @ 2'} = \frac{870}{12 \times 24} \pm \frac{459 \times 12}{1152}$$

     = 3.02 ± 4.78
Max. comp. = 7.8 max. tension = 1.76   OK

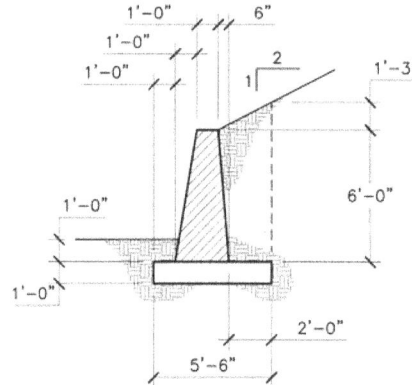

Resisting Moments:

			W		X		M	
Stem:								
1 x 6 x 145	=		870	x	2.50	=	2,175	ft. lbs.
+ 1 x 6 x 0.5 x 145	=		435	x	1.67	=	726	
+ 0.5 x 6 x 0.5 x 145	=		218	x	3.17	=	691	
Soil:								
1 x 1 x 110	=		110	x	0.50	=	55	
+ 0. 5 x6 x 0.5 x 110	=		165	x	3.33	=	549	
+ 2 x 6 x 110	=		1,320	x	4.50	=	5,940	
+ 1.25 x 2.5 x 0.5 x 110 =			172	x	4.67	=	803	
Footing:								
1 x 5.5 x 150	=		825	x	2.75	=	2,269	
			4,115	lbs.			3,208	ft.-lbs.

Overturning M $= \dfrac{43\,(1.25 + 6.0 + 1.0)^3}{2 \times 3} - \dfrac{30 \times 2^3}{2 \times 3} = 3944$ ft – lbs.

Overturning S. F. $= \dfrac{13,208}{3944}$ 3.34 > 1.5 OK

Dist. toe to soil pressure c.g.

$$\dfrac{13,208 - 3944}{4115} = 2.25 \text{ ft.}$$

DESIGN EXAMPLE 9 $\qquad$ **Page 3 of 3**

$$e = \frac{5.50}{2} - 2.25 = 0.50$$

within middle third

Soil pressure $\quad = \dfrac{4115}{5.5} \pm \dfrac{6 \, x \, 4115 \, x \, 0.50}{5.5^2}$

$\qquad\qquad = 748 + 408 = 1156 \, psf < 3500 \;\; OK$

Check sliding:

$\qquad$ Net lateral force $= \dfrac{43 \, (1.25 + 6.0 + 1.0)^2}{2} = 1363 \, lbs.$

$\qquad$ Cohesion resistance $= 5.5 \, x \, 200 = 1100 \, lbs.$

$\qquad$ Passive resistance $= \dfrac{300 \, x \, 2^2}{2} = 600$

$\qquad$ Sliding S.F. $= \dfrac{600 + 1100}{1363} = 1.17 < 1.5 \;\; Consider \; key$

DESIGN EXAMPLE 9   Report Printout

Basics of Retaining Wall Design

Title EX-9
Job #
Description ...
EX-9

Dsgnr: HB   Date: 2 JUL 2018   Page: 1

This Wall in File: c:\Users\chris\Dropbox\RetainPro 10 Project Files\HB\Hugh.RPX

RetainPro (c) 1987-2018, Build 11.18.06.22
License : KW-06000216
License To : RetainPro Development Center

Gravity Stem Retaining Wall   Code: IBC 2015,ACI 318-14,ACI 530-13

Criteria

Retained Height	=	6.00 ft
Wall height above soil	=	0.00 ft
Slope Behind Wall	=	2.00
Height of Soil over Toe	=	12.00 in
Soil Density	=	110.00 pcf

Soil Data

Allow Soil Bearing	=	2,000.0 psf
Equivalent Fluid Pressure Method		
Heel Active Pressure	=	43.0 psf/ft
Passive Pressure		300.0 psf/ft
Water height over heel	=	0.0 ft
Cohesion value	=	200.0 psf
Soil height to ignore for passive pressure	=	0.00 in

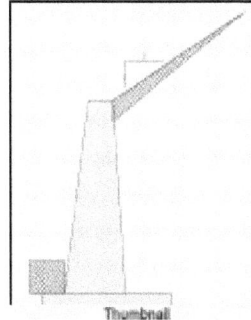

Thumbnail

Surcharge Loads

Surcharge Over Heel	=	0.0 psf
>>>Used To Resist Sliding & Overturning		
Surcharge Over Toe	=	0.0 psf
Used for Sliding & Overturning		

Lateral Load Applied to Stem

Lateral Load	=	0.0 #/ft
Height to Top	=	0.00 ft
Height to Bottom	=	0.00 ft
Wind on Exposed Stem	=	0.0 psf

Adjacent Footing Load

Adjacent Footing Load	=	0.0 lbs
Footing Width	=	0.00 ft
Eccentricity	=	0.00 in
Wall to Ftg CL Dist	=	0.00 ft
Footing Type		Line Load
Base Above/Below Soil at Back of Wall	=	0.0 ft
Poisson's Ratio	=	0.300

Design Summary

Total Bearing Load	=	4,124 lbs
...resultant ecc.	=	6.26 in
Soil Pressure @ Toe	=	1,176 psf OK
Soil Pressure @ Heel	=	323 psf OK
Allowable	=	2,000 psf
Soil Pressure Less Than Allowable		
ACI Factored @ Toe	=	1,647 psf
ACI Factored @ Heel	=	453 psf
Footing Shear @ Toe	=	9.9 psi OK
Footing Shear @ Heel	=	8.1 psi OK
Allowable	=	67.1 psi
Sliding Stability Ratio	=	1.16 Ratio < 1.5

Sliding Calcs (Vertical Component NOT Used)

Lateral Sliding Force	=	1,463.3 lbs
less Passive Force	= -	600.0 lbs
less Cohesion Force	= -	1,100.0 lbs
Added Force Req'd	=	0.0 lbs OK
...for 1.5 Stability	=	0.0 lbs OK

Load Factors

Building Code	IBC 2015,ACI
Dead Load	1.200
Live Load	1.600
Earth, H	1.600
Wind, W	1.000
Seismic, E	1.000

Rubble masonry, mortar bonded Stem Analysis Data *(Unreinforced material)*

Wall Material Weight	=	145.00 pcf
Fc: Max. Allow. Compression	=	100.0 psi
Fc: Max. Allow. Tension	=	10.0 psi
Front Batter Distance	=	12.00 in
Thickness @ Top of Stem	=	12.00 in
Back Batter Distance	=	6.00 in

		@ Height #1	@ Height #2	@ Height #3
		OK	OK	Tension Exceeded
Height above Footing	=	4.00 ft	2.00 ft	0.00 ft
Wall Thick. @ Height	=	18.00 in	24.00 in	30.00 in
Section Modulus	=	648.00 in^3	1,152.00 in^3	1,800.00 in^3
Moment @ Height	=	91.7 ft-#	733.9 ft-#	2,476.8 ft-#
Vertical Load @ Height	=	362.5 lbs	870.0 lbs	1,522.5 lbs
Actual Unit Tension	=	0.0 psi	4.6 psi	12.3 psi
Actual Unit Compression	=	3.4 psi	10.7 psi	20.7 psi
Shear @ Section	=	137.6 lbs	550.4 lbs	1,238.4 lbs
Actual Unit Shear	=	0.0 psi	0.0 psi	0.0 psi

DESIGN EXAMPLE 9 Report Printout

Basics of Retaining Wall Design

Title: EX-9
Job #
Description: EX-9
Dsgnr: HB
Date: 2 JUL 2016
Page: 2

This Wall in File: c:\Users\chris\Dropbox\RetainPro 10 Project Files\HB\Hugh.RPX

RetainPro (c) 1987-2016, Build 11.16.06.22
License : KW-06006216
License To : RetainPro Development Center

Gravity Stem Retaining Wall
Code: IBC 2015,ACI 318-14,ACI 530-13

Footing Strengths & Dimensions

Toe Width	=	1.00 ft
Heel Width	=	4.50
Total Footing Width	=	5.50
Footing Thickness	=	12.00 in
Key Width	=	0.00 in
Key Depth	=	0.00 in
Key Distance from Toe	=	0.00 ft
f'c = 2,000 psi	Fy =	60,000 psi
Footing Concrete Density	=	150.00 pcf
Min. As %	=	0.00*8
Cover @ Top = 2.00 in	@ Btm =	3.00 in

Footing Design Results

		Toe	Heel
Factored Pressure	=	1,647	463 psf
Mu' : Upward	=	787	1,195 ft-#
Mu' : Downward	=	351	3,418 ft-#
Mu : Design	=	436	2,223 ft-#
Actual 1-Way Shear	=	9.65	6.14 psi
Allow 1-Way Shear	=	35.78	35.78 psi
Toe Reinforcing	=	#7 @ 18.00 in	
Heel Reinforcing	=	#6 @ 16.00 in	
Key Reinforcing	=	None Spec'd	

Other Acceptable Sizes & Spacings

Toe: Not req'd, Mu < phi*5*lambda*sqrt(fc)*Sm
Heel: Not req'd, Mu < phi*5*lambda*sqrt(fc)*Sm
Key: No key defined

Summary of Overturning & Resisting Forces & Moments

		....OVERTURNING....					...RESISTING...		
Item		Force lbs	Distance ft	Moment ft-#			Force lbs	Distance ft	Moment ft-#
Heel Active Pressure	=	1,463.3	2.75	4,024.2	Soil Over Heel	=	1,320.0	4.50	5,940.0
Surcharge Over Toe	=				Sloped Soil Over Heel	=	171.9	4.67	802.1
Adjacent Footing Load	=				Surcharge Over Heel	=			
Added Lateral Load	=				Adjacent Footing Load	=			
Load @ Stem Above Soil	=				Axial Dead Load on Stem	=		0.00	
Seismic Load	=				Soil Over Toe	=	119.2	0.54	64.7
Seismic Stem Self Wt	=				Surcharge Over Toe	=			
					Stem Weight	=	1,522.5	2.36	3,588.8
Total	=	1,463.3	O.T.M.	4,024.2	Earth above Sloping Stem	=	166.0	3.33	550.0
					Footing Weight	=	825.0	2.75	2,268.8
Resisting/Overturning Ratio	=		3.28		Key Weight	=			
Vertical Loads used for Soil Pressure	=		4,123.5 lbs		Vert. Component	=			
Vertical component of active pressure NOT used for soil pressure					Total =		4,123.5 lbs	R.M.=	13,214.3

Tilt

Horizontal Deflection at Top of Wall due to settlement of soil

(Deflection due to wall bending not considered)

Soil Spring Reaction Modulus	250.0 pci
Horizontal Defl @ Top of Wall (approximate only)	0.036 in

The above calculation is not valid if the heel soil bearing pressure exceeds that of the toe, because the wall would then tend to rotate into the retained soil.

DESIGN EXAMPLE 10   **Page 1 of 8**

Segmental Retaining Wall – Geogrids

Wall height = 12.0 ft.

Embedment = 1.0 ft.

Backfill slope = 4:1 – 14°

Backfill soil: $\phi_b = 33°$ $\gamma = 120$ pcf

   $\delta = \frac{2}{3}\phi = 22°$

In situ soil: $\phi_i = 32°$ $\gamma = 110$ pcf

   $\delta = \phi_i = 32°$

Use Coulomb method

DL = 50 psf   LL = 100 psf

Base width (trial) = 75% of 12 ft. say

   = 10.00 ft.

Block selection: Keystone Compac

   1" offsets, Height = 8.0 in.,

   Depth = 12.0 in.

      Wt. = 120 psf.

   Batter = 7.1° $\alpha = 90 + 7.1 = 97.1°$

Active earth pressure – backfill zone

$$K_a = \frac{\sin^2(\alpha + \phi_i)}{\sin^2\alpha\,\sin(\alpha - \delta)\left[1 + \sqrt{\dfrac{\sin(\phi + \delta)\sin(\phi - \beta)}{\sin(\alpha - \delta)\sin(\alpha - \beta)}}\right]^2}$$

$$= \frac{\sin^2(97.1 + 33)}{\sin^2 97.1\,\sin(97.1 - 22)\left[1 + \sqrt{\dfrac{\sin(33 + 22)\sin(33 - 14.0)}{\sin(97.1 - 22)\sin(97.1 - 14.0)}}\right]^2}$$

   = 0.26

K_a (horiz) = 0.26 cos (22 - 7.1) = 0.25

Geogrid Placement:

Lowest layer at 1st block joint = +0.67 ft.

Space subsequent layers every 2nd block = 1.33 ft. o.c.

DESIGN EXAMPLE 10   **Page 2 of 8**

Tension to Bottom Layer:

$$T_u \;=\; 0.25 \times 120 \left(\frac{2.0 - 0}{2} \right) \times (12.0 - 0.67) \qquad = 340$$

$$+ \text{ surcharge: } 0.25\,(50 + 100) \times \left(\frac{2.0 - 0}{2} \right) \qquad = \underline{38}$$

(Sloped backfill neglected) T_u =   378   lbs.

Select Geogrid

Try Strata Systems Stratagrid SG200

Long-term design strength (LTDS) = 1613

S.F. = 1.5

Allowable (LTADS) $= \dfrac{1643}{1.5} = 1085 > 378$ @ bott. layer   OK

Check Connection Strength

Equations: Peak connect $= 889 + 0.31\,N$ but < 1624

           ¾" Serviceability $= 519 + 0.14\,N$ but < 767

$N \;=\; (12.0 - 0.67)\,120 \;=\; 1360$ lbs.

Peak connect value $= \dfrac{(889 + .31 \times 1360)}{1.5} = 874 > 378$ OK

¾" Serviceability $= \dfrac{(519 + 0.14 \times 1360)}{1.0} = 709 > 378$ OK

Safety Factor   =

$$\frac{Lesser\ of\ Peak, Service, or\ LTADS}{T_u} = \frac{709}{378} = 1.88 > 1.50 \quad \text{OK}$$

Check Embedment Depth, L_e

Bottom Layer:

$$L_e \;=\; \frac{T_u}{2 H_{ov} \tan \phi_i \times C_i} \qquad (H_{ov} = \text{overlay soil} + \text{surcharge})$$

Assume $C_i = 0.90$ $H_{ov} = (12.0 - 0.67)\,120 + (50 + 100) = 1510$ lbs.

DESIGN EXAMPLE 10   **Page 3 of 8**

$$L_e = \frac{378}{2 \; x \; 1510 \; x \; \tan 33° \; x \; 0.90} = 0.22 \text{ ft.}$$

Add 1.0 ft. per NCMA = 1.22 ft. < 7.00 (AASHTO requires 3.0 ft. added = 3.22 ft.)

Tension to Layer #2:

$$T_u = 0.25 \; x \; 120 \left(\frac{3.33 - 0.67}{2}\right) x \; (12.0 - 2.0) \qquad = \qquad\qquad 399$$

$$+ \text{ surcharge: } 0.25 (50 + 100) \; x \left(\frac{3.33 - 0.67}{2}\right) \qquad = \qquad\qquad \underline{50}$$

(Sloped backfill neglected) T_u = 449 lbs.

Check Connection Strength

Equations: Peak connect   = 889 + 0.31 N but < 1624

               ¾" Serviceability   = 519 + 0.14 N but < 767

N = (12.0 – 2.0) 120 = 1200 lbs.

Peak connect value $= \dfrac{(889 + .31 \; x \; 1200)}{1.5} = 841 > 449 \qquad$ OK

¾" Serviceability $= \dfrac{(519 + 0.14 \; x \; 1200)}{1.0} = 687 > 449 \qquad$ OK

Tension to Top Layer:

$$T_u = 0.25 \; x \; 120 \left(\frac{12.0 - 10.0}{2}\right) x \; (12.0 - 11.33) \qquad = \qquad\qquad 20$$

$$+ \text{ surcharge: } 0.25 (50 + 100) \; x \left(\frac{12.0 - 10.0}{2}\right) \qquad = \qquad\qquad \underline{38}$$

(Sloped backfill neglected) T_u =   58   lbs.

Check Geogrid

Strata Systems Stratagrid SG200

Long-term design strength (LTDS) = 1643

S.F. = 1.5

Allowable (LTADS) $= \dfrac{1643}{1.5} = 1095 > 58$ OK

DESIGN EXAMPLE 10   **Page 4 of 8**

Check Connection Strength

Equations: Peak connect $= 889 + 0.31$ N but < 1624

                ¾" Serviceability $= 519 + 0.14$ N but < 767

N = $(12.0 - 11.33)\ 120 = 80$ lbs.

Peak connect value $= \dfrac{(889 + .31\ x\ 80)}{1.5} = 609 > 58$ OK

¾" Serviceability $= \dfrac{(519 + 0.14\ x\ 80)}{1.0} = 530 > 58$ OK

Check Embedment Depth, L_e

$$L_e = \frac{T_u}{2H_{ov}\tan\phi_i\ x\ C_i}$$ (H_{ov} = overlay soil + surcharge)

Assume $C_i = 0.90$ $H_{ov} = (12.0 - 11.33)\ 120 + (50 + 100) = 230.4$ lbs.

$$L_e = \frac{58}{2\ x\ 230\ x\ \tan 33°\ x\ 0.90} = 0.22$$

Add 1.0 ft. per NCMA = 1.22 ft   (AASHTO requires 3.0 ft. added = 3.22 ft.)

Check available embedment depth based on base = 10.0 ft.

Coulomb rupture angle = 52.1°

L_e avail.: $(10.0 - 1.0) + 11.33 \tan 7.1° - \dfrac{11.33}{\tan 52.1} = 1.59 > 1.22$ OK

Overturning Moments

NOTE: Earth pressure applied to back of reinforced zone, assuming Vertical Plane (90°)
        effective ht. = $12.00 + 9.00 \tan 14 = 14.25$ ft.

K_a [external (in-situ)]

    ϕ_e = 32°
    α = 90°
    δ = $\phi = 32°$
    β = 14°
    γ = 110

K_a = 0.34

K_c (horiz) = 0.29

	Force	Distance	Moment

Earth pressure $0.29 \times 110 \times \dfrac{14.25^2}{2}$ =   3239 $\dfrac{14.25}{3}$ = 4.75   15,385   ft.-lbs.

Surcharge $0.29 (50 + 100) \times 14.25$ =   620 $\dfrac{14.25}{2}$ = 7.1   4,418

Sliding Force Total =   3859   lbs.   19,803   ft. lbs.

<u>Resisting Moments</u>

		Force	Distance	Moment
W_4 Wall	12×120 =	1440	1.25	1,800
W_1 Earth	$9 \times 12 \times 120$ =	12,960	5.5	71,280
W_2 Sloped	$9 \times 2.25 \times 120 \times \frac{1}{2}$ = 1215		$\frac{2}{3} \times 9 + 2.49 = 8.5$	10,328
W_3 Surcharge	$9 \times (50 + 100)$ = 1350		$\frac{9}{2} + 2.49 =$ 7.0	9,436
Total vert. force =		16,965 lbs.		92,844 ft. lbs.

Overturning safety factor ratio = $\dfrac{92,844}{19,803}$ = 4.68 > 2.0   OK

<u>Check Sliding at Base</u>

Lateral force on reinforced soil = 3859 lbs.

Sliding resistance = 16,965 tan ϕ_e = 10,600 lbs.

Total vertical force = 16,965 lbs.

Sliding safety factor = $\dfrac{10,600}{3859}$ = 2.75   OK

<u>Check Soil Pressure</u>

Use Meyerhoff Method

Eccentricity, e = $\dfrac{(B)}{2} - \dfrac{M_R - M_{OT}}{V_T} = \dfrac{(10)}{2} - \dfrac{92,84 - 19,803}{16,965} = 5.0 - 4.31 = 0.69$ *ft.*

Effective bearing length = (B) − 2e = (10) − 2 × 0.69 = 8.62 ft.

DESIGN EXAMPLE 10   **Page 6 of 8**

Bearing pressure = $\dfrac{V_T}{8.62} = \dfrac{16,965}{8.62}$ = 1968 psf

Allowable Bearing Pressure

Assume no cohesion (c = 0)

   = $\gamma D N_q + 0.5\gamma$ [Eff. Bearing length] N_γ γ = in situ soil density = 110 pcf.

   = 110 x 1.0 x 23.2 + .5 x 110 (8.62) 30.22   D = Depth of embedment

      = 1.0 ft.

   = 16,879 psf N_q for 32° = 23.2

                                                 N_γ for 32° = 30.22

Soil bearing ratio   = $\dfrac{16,879}{1968}$ = 8.58   OK

Values for N_q and N_γ

ϕ_i	N_q	N_γ
31	20.63	26.0
32	23.2	30.2
33	26.1	35.2
34	29.4	41.1
35	33.3	48.0
36	37.8	56.3

Check for Added Seismic

Added seismic has three components:

   Seismic force of self-weight of wall

   Seismic force from reinforced zone

   Seismic force acting on reinforced zone

k_h =   0.15 θ = $\tan^{-1} 0.15 = 8.53°$

K_{AEH} =   0.55 [α = 97.1°, ϕ_i = 33°, δ = 22°, β = 14°]

K_{AH} = 0.25 $\Delta K_{AEH} = K_{AEH} - K_{AH}$ = .55 - .25 = 0.30

DESIGN EXAMPLE 10   **Page 7 of 8**

Total seismic lateral force:

Wall: K_h x W x H = 0.15 x 120 x 12   =   216 lbs.

Reinf. Zone: K_h x H (.5 H – t) γ_i = 0.15 x 12 (.5 x 12 – 1.0) 120   =   1080 lbs.

   + sloped soil = k_h [(.5 x 12 – 1.0)^2 x γ_i x tan β x 0.5]

   = .15 [(.5 x 12 – 1.0)^2 x 120 x tan 14° x .5]   =   56

* Exterior of zone: = 0.5 x ΔK_{AEH} [H + (B-t) tan β x 0.5]^2 x 0.5 γ_e

   = .5 x 0.30 [12 + (10 – 1) tan 14°]^2 x 0.5 x 110   =   <u>1675</u> lbs.

* Use 50% of seismic per NCMA

                                Total seismic   =   3027 lbs.

** Total sliding force   = 0.5 K_{AH} [H + (B – t) tanβ]^2 γ_e + seismic

   = 0.5 x 0.29 x [12 + (10-1) tan 14°]^2 x 110 + 3027 =   6266 lbs.

** K_{AH} for back of reinf. zone based on α= 90° and δ = ϕ_e ∴ K_{AH} =   0.29

+ (DL + LL) = 0.29 (50 + 100) x [12 + (10 – 1) tan 14°]   =   620

Total sliding   = 6266 + 620 = 6886 lbs.

Sliding resistance   = 10,660 lbs.

Sliding ratio $= \dfrac{10,660}{6886} = 1.55$ OK

<u>Added Seismic Overturning</u>

Wall: OTM $= 216 \times \dfrac{12}{2}$ =   1296

Reinf. Zone $= 1080 \times \dfrac{12}{2}$ =   6480

Exterior: = 1675 x [12 + (10 - 1) tanβ] x 0.6   = 14,321

56 x [12 + (.5H tan 14° x 0.33)]   =   700

Total added seismic overturning   = 22,797 ft. lbs.

DESIGN EXAMPLE 10

Total overturning = 22,799 + 19,803 = 42,782 ft. lbs.

Overturning ratio w/seismic = $\dfrac{92,852}{42,782}$ = 2.17

Seismic Tension to Layer #1

$$k_h\left(\frac{h_2-h_o}{2}\right)w + \Delta K_{AEH}\,\gamma_i\,H\left(\frac{h_2-h_o}{2}\right)\left[0.8-0.6\left(\frac{H-h_o}{H}\right)\right]$$

$$= 0.15\left(\frac{2.0-0}{2}\right)120 + 0.30 \times 120 \times 12\left(\frac{2.0-0}{2}\right)\left[0.8-0.6\left(\frac{12-0.67}{12}\right)\right]$$

$$= 119 \text{ lbs.}$$

Pullout safety factor = $\dfrac{1064}{378+119}$ = 2.14

Seismic tension to top layer #9

$$= 0.15\left(\frac{12-10}{2}\right)120 + 0.30 \times 120 \times 12\left(\frac{12-10}{2}\right)\left[0.8-0.6\left(\frac{11.33}{12}\right)\right]$$

$$= 349 \text{ lbs.}$$

Pullout S.F. = $\dfrac{540}{58+349}$ = 1.28 > 1.1 (for seismic) OK

$$L_e = \frac{58+349}{2\,x\,[(0.67+say\,2.0)\,x\,120+(50+100)]\tan 33° \,x\,0.90} + 1.0 = 1.66 \text{ ft.}$$

Available embedment = (10-1) + 11.33 tan 7.1° - $\dfrac{11.33}{\tan 52.1°}$ = 1.59

Consider OK w/seismic

DESIGN EXAMPLE 10 — Report Printout

Basics of Retaining Wall Design

Title EX-10
Job #
Description
EX-10

Dsgnr: HB

Date: 2 JUL 2018

Page: 1

This Wall in File: c:\Users\christ\Dropbox\RetainPro 10 Project Files\HB\Hugh.RPX

RetainPro (c) 1987-2018, Build 11.18.08.22
License : KW-06006216
License To : RetainPro Development Center

Segmental Retaining Wall with Geogrids

Code: NCMA

Criteria

Wall height (retained height), ft	12.00
Backfill slope	4:1
Backfill angle	14.0
Embedment	1.0

Soil data

External Soil, Phi_e	33
External soil density (in situ), pcf	110
Internal Soil, Phi_i	33
Internal soil density, pcf	120
Wall Soil Friction Angle	22
K_a(Horiz)	0.26

Stability

Overturning ratio	4.98
Sliding ratio	3.03
Overturning moment, ft-lbs	18,650
Resisting moment, ft-lbs	92,866
Total lateral/sliding force, lbs	3,635
Sliding Resistance, ft	11,015.08
Total vertical force, lbs	16,962
Base length, ft	10.00
Eccentricity on base, ft	0.62
Effective base length, ft	8.75
Soil bearing pressure, psf	1,938.27
Allowable soil bearing, psf	16,902.46
Soil Bearing Ratio	8.72

Loading

Dead load, psf	50
Live load, psf	100
Seismic Design Kh	0.00

Segmental block data

Vendor selection	'Keystone Retaining Wall Systems'
Vendor web address	'www.keystonewalls.com'
Block selection type	'Compac'
Block height, in	8.00
Block depth, in	12.00
Offset per block, in	1.00
Batter angle	7.13
Wall weight, psf	120.00
Hinge height, ft	8.00

Geogrid material

Vendor Selection	'Strata Systems, Inc.'
Vendor web address	'www.geogrid.com'
Geogrid type	'Stratagrid SG200'
LTDS	1689.15
Factor of safety	1.50
LTADS	1126.10
Peak connection equation	889 + 0.31N
Peak connection maximum	1624
Serviceability connection equation	519 + 0.14N
Serviceability connection maximum	767

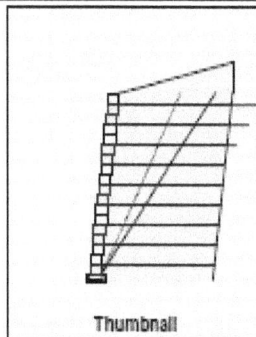

Wall Analysis Table:

Block	Layer	Height above base Ft In		Dec	Tension Static	Seismic	Connection Peak	Serv	Embed Le	Vert N	S.F.	Required Extent, Ft
18		12'	0"	12.00								
17	9	11'	4"	11.33	57		609.2	530.2	1.57	80	0.26	9.96
15	8	10'	0"	10.00	130		642.3	552.6	1.57	240	4.27	9.10
13	7	8'	8"	8.67	183		675.3	575.0	1.52	400	3.15	8.17
11	6	7'	4"	7.33	236		708.4	597.4	1.50	560	2.53	7.28
9	5	6'	0"	6.00	289		741.5	619.8	1.48	720	2.14	6.40
7	4	4'	8"	4.67	342		774.5	642.2	1.47	880	1.88	5.52
5	3	3'	4"	3.33	395		807.6	664.6	1.47	1,040	1.68	4.64
3	2	2'	0"	2.00	448		840.7	687.0	1.48	1,200	1.53	3.76
1	1	0'	8"	0.67	376		873.7	709.4	1.34	1,360	1.90	2.88
Base		0'	0"	0.00						1,440		

DESIGN EXAMPLE 10   Report Printout

Basics of Retaining Wall Design

Title: EX-10
Job #
Description:
EX-10

Dsgnr: HB

Date: 2 JUL 2018

Page: 2

This Wall in File: c:\Users\chris\Dropbox\RetainPro 10 Project Files\HB\Hugh.RPX

RetainPro (c) 1987-2018, Build 11.18.06.22
License : KW-06006216
License To : RetainPro Development Center

Segmental Retaining Wall with Geogrids

Code: NCMA

* Extent of geogrid referenced from the front face of wall. FT

Summary: Resisting / Overturning

Resisting Moments

Item	Force, lbs	Distance, ft	Moment, ft-lbs
Wall	1,440	1.25	1,800
Reinf. earth	12,960	5.50	71,280
Sloped	1,212	8.53	10,336
Dead load	450	7.00	3,150
Live load	900	7.00	6,300
Total	16,962		92,866

Overturning Moments

Item	Force, lbs	Distance, ft	Moment, ft-lbs
Earth	3,049	4.75	14,475
Surcharge, DL	195	7.12	1,392
Surcharge, LL	391	7.12	2,784
Seismic, Wall	0	0.00	0
Seismic, Reinf	0	0.00	0
Seismic, Sloped soil	0	0.00	0
Seismic, Exterior	0	0.00	0
Total	3,635		18,650

Overturning Ratio   4.98

ASSUMPTIONS AND CRITERIA USED

1. References used include *Design Manual for Segmental Retaining Walls, 2nd Edition*, and *Segmental Retaining Walls – Seismic Design Manual, 1st Edition*, both by NCMA.
2. Blocks are all same size and uniform offsets (batter) for full wall height.
3. Coulomb earth pressure theory used for earth pressures and failure plane angle.
4. Refer to geotechnical report for backfill material, compaction, and other design data and recommendations.
5. Cap blocks if used are above the retained height and are neglected in this design.
6. Geogrid LTDS and connection values for block vendors obtained from ICC Evaluation Service (ES Legacy Reports) or as provided by vendors. Since these may change or be updated, verification of values is recommended.
7. Block sizes obtained from vendors' literature and may vary with locality.
8. Geogrid layers are equally spaced vertically, all same length, and laid horizontally.
9. Average weight of block and cell infill assumed to be 120 pcf.
10. See vendor web sites (on input screen) for more information and specifications.
11. Design height is limited to 16 feet or 24 blocks, whichever is less. Contact vendor for higher designs or special conditions.
12. Seismic design is per Seismic Design Manual cited above. Also see Methodology/Seismic Design in User's Manual.
13. Vendor specifications or project specifications, whichever is most restrictive, to be followed for construction procedures.
14. Add notes and details for proper drainage.
15. See User's Manual Design Example #10 for methodology and sample verification calculations.
16. Final design responsibility is with the project Engineer-of-Record.

DESIGN EXAMPLE 11   **Page 1 of 3**

Segmental Gravity Wall Design

Criteria:

Retained height (trial) = 4.0 ft.

Slope: Level

Soil density, in situ – 110 pcf

Soil density, backfill – 120 pcf

Soil ϕ, backfill 33°

Soil ϕ, in-situ 34°

Soil/wall friction angle = ⅔ϕ backfill

$\quad$ = 22°

Embedment = 1.0 ft.

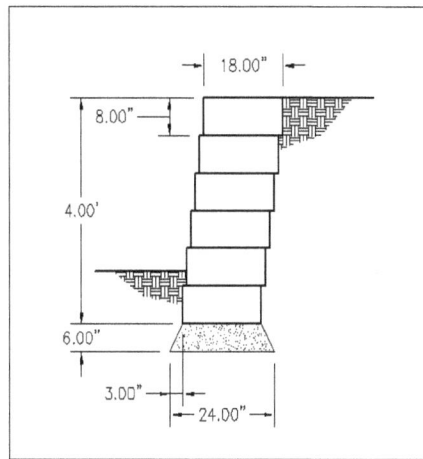

Block Data:

Try Keystone, Standard

Height = 8.00 in.   Depth = 18.0"

Wt. = 120 x 1.5 180 psf

Offset: ½ in. per block

Batter = $\tan^{-1}$ (0.5/8.0) = 3.6° = ω

Hinge height = $\dfrac{(blockdepth)}{\tan \omega} = \dfrac{1.50}{\tan 3.6°} = \dfrac{1.50}{.063} = 24\ ft. > 4.0\ ft.$

Lateral Soil Pressure:

$$K_a = \frac{\sin^2 (\alpha + \varphi)}{\sin^2 \alpha\, \sin (\alpha - \delta) \left[1 + \sqrt{\dfrac{\sin (\varphi + \delta)\, \sin (\varphi - \beta)}{\sin (\alpha - \delta)\, \sin (\alpha - \beta)}} \right]^2}$$

K_a (horiz.) = $K_a \cos (90 + \delta - \alpha)$

$$\frac{\sin^2 (93.60 + 33.0)}{\sin^2 93.60\, \sin (93.60 - 22.0) \left[1 + \sqrt{\dfrac{\sin (33.0 + 22.0)\, \sin (33.0 - 0)}{\sin (93.60 - 22.0)\, \sin (93.60 - 0)}} \right]^2} = 0.24$$

K_a (horiz.) = 0.24 $\cos (90 + 22.0 - 93.60) = 0.23$

Total lateral force = $\dfrac{0.23\, x\, 120\, x\, 4^2}{2} = 221\ lbs.$

DESIGN EXAMPLE 11	Page 2 of 3

Stability:

Overturning moment = $221 \times \dfrac{4.0}{3}$ = 294 ft. lbs.

Total vertical force = 4.0 x 180 pcf = 720 lbs.

Resisting moment = $720 \left[\left(\dfrac{1.5}{2} \right) + \left(\dfrac{4.0}{2} \tan 3.6° \right) \right]$ = 630 ft. lbs.

Overturning ratio = $\dfrac{630}{294}$ = 2.14 > 1.50   OK

Check Sliding

Sliding force = 221 lbs.

Resístanse = 720 (tan ϕ in situ) = 720 x tan 34° = 486 lbs.

Sliding safety factor = $\dfrac{486}{221}$ = 2.20 > 2.0   OK

Soil Bearing Pressure

Base width = 1.0 + 0.5 = 1.5 ft. = B

$e = \dfrac{B}{2} - \left(\dfrac{Resisting - Overturning}{Total\ Vert.} \right) = \dfrac{1.5}{2} - \dfrac{630 - 294}{720} = 0.28$

Effective bearing area = (B – 2e) = [(1.5 – 2 x 0.28) +0.50] = 1.44 ft.

Bearing pressure = $\dfrac{720}{1.44}$ = 500 psf.

Allowable Bearing Capacity:

Effective bearing area = 1.44 ft.

Depth to bottom of 6 in. pad = 1.0 + 0.5 = 1.5 ft.

$N_q = 29.4$ $N_\gamma = 41.1$

Bearing capacity = $\gamma D N_q + 0.5\gamma (B – 2_e) N_\gamma$ (N_q and N_y from Table in NCMA Handbook)

   = 110 x 1.5 x 29.4 + 0.5 x 110 x [(1.5 – 2 x 0.28) + 0.5] 41.1

   = 8106 psf   OK

Bearing safety factor = $\dfrac{8106}{500}$ = 16.2

DESIGN EXAMPLE 11   **Page 3 of 3**

If Design for Seismic:

Assume $k_h = 0.05$ (If greater, overturning would exceed resisting moment!)

K_a (horiz.) = 0.23

$$K_{AE} = \frac{\sin^2 (\alpha + \theta - \phi)}{\cos \theta \sin^2 \alpha \sin (\alpha + \theta + \delta) \left[1 + \sqrt{\dfrac{\sin (\phi + \delta) \sin (\phi - \beta - \theta)}{\sin (\alpha + \theta + \delta) \sin (\alpha - \beta)}}\right]^2}$$

$\theta = \tan^{-1} (0.05) = 2.9°$

$$= \frac{\sin^2 (93.60 + 2.9 - 33.0)}{\cos 2.9 \sin^2 93.60 \sin (93.60 + 2.9 + 22.0) \left[1 + \sqrt{\dfrac{\sin (33.0 + 22.0) \sin (33.0 - 0.0 - 2.9)}{\sin (93.60 + 2.9 + 22.0) \sin (93.60 - 0.0)}}\right]^2}$$

K_{AE} (horiz.) = 0.32 cos (90 + 22.0 − 93.60) = 0.30

$K_{AH} = 0.23$

$\Delta K_{AEH} = 0.30 - 0.23 = 0.07$

Added seismic = 0.07 $(4)^2$ x 120 x ½ = 67 lbs.   + 0.05 x 4.0 x 180 = 36 lbs.

Total sliding = 221 + 67 + 36 = 324 lbs.

Overturning moment = 294 + 67 x 0.60 x 4 + 36 x 2 = 527 ft. lbs.

Overturning ratio = $\dfrac{630}{547}$ = 1.2 < 1.1

(Note: Safety factor when seismic included = 0.75 x 1.5 = 1.1 per IBC '15)

This design example used a very low seismic factor for illustration. A higher seismic factor would require a revised design.

DESIGN EXAMPLE 11   **Report Printout**

Basics of Retaining Wall Design   Title   EX-11   Page   1
                                                   Job #   Dsgnr   HB   Date   2 JUL 2018
                                                   Description ...
                                                   EX-11

This Wall in File: c:\Users\chris\Dropbox\RetainPro 10 Project Files\HB\Hugh.RPX

RetainPro (c) 1987-2018, Build 11.18.08.22
License : KW-06000018   **Segmental Gravity Retaining Wall**   Code: NCMA
License To : RetainPro Development Center

Criteria

Wall height (retained height), ft	4.00
Backfill slope	Level
Backfill angle	0.0
Embedment	1.0

Soil data

External Soil, Phi_e	34
External soil density (In situ), pcf	110
Internal Soil, Phi_i	33
Internal soil density, pcf	120
Wall Soil Friction Angle	22
K_a(Horiz)	0.21

Stability

Overturning ratio	2.68
Sliding ratio	2.41
Overturning moment, ft-lbs	269
Resisting moment, ft-lbs	720
Total lateral/sliding force, lbs	201
Sliding Resistance, ft	485.65
Total vertical force, lbs	720
Base length, ft	1.50
Eccentricity on base, ft	0.12
Effective base length, ft	1.75
Soil bearing pressure, psf	410.54
Allowable soil bearing, psf	8,107.00
Soil Bearing Ratio	19.75

Segmental block data

Vendor selection	'Keystone Retaining Wall Systems'
Vendor web address	'www.keystonewalls.com'
Block selection type	'Standard'
Block height, in	8.00
Block depth, in	18.00
Offset per block, in	1.00
Batter angle	7.13
Wall weight, psf	180.00
Hinge height, ft	12.00

Thumbnail

Loading

Dead load, psf	0
Live load, psf	0
Seismic Design Kh	0.00

Wall Analysis Table:

Block	Height above base Ft In	Height above base Dec	Vert N	Lateral Static	Lateral Seismic	Shear Interface	S.F.
6	4' 0"	4.00				1,550.00	
5	3' 4"	3.33	120	8		1,587.20	283.59
4	2' 8"	2.67	240	22		1,624.40	72.56
3	2' 0"	2.00	360	50		1,661.60	32.99
2	1' 4"	1.33	480	90		1,698.80	18.97
1	0' 8"	0.67	600	140		1,736.00	12.41
Base	0' 0"	0.00	720	201		1,773.20	8.80

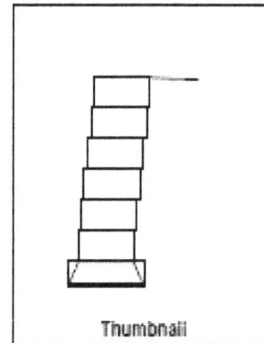

DESIGN EXAMPLE 11　　　　　　　　　　　**Report Printout**

Basics of Retaining Wall Design

Title　EX-11
Job #:
Description
EX-11

Dsgnr　HB

Date:　2 JUL 2018

Page　2

This Wall in File: c:\Users\chris\Dropbox\RetainPro 10 Project Files\HB\Hugh.RPX

RetainPro (c) 1987-2018, Build 11.18.06.22
License : KW-06000216
License To : RetainPro Development Center

Segmental Gravity Retaining Wall

Code: NCMA

ASSUMPTIONS AND CRITERIA USED

1. References used include *Design Manual for Segmental Retaining Walls, 2nd Edition,* and *Segmental Retaining Walls – Seismic Design Manual, 1st Edition,* both by NCMA.
2. Blocks are all same size and uniform offsets (batter) for full wall height.
3. Coulomb earth pressure theory used for earth pressures and failure plane angle.
4. Refer to geotechnical report for backfill material, compaction, and other design data and recommendations.
5. Cap blocks if used are above the retained height and neglected in this design.
6. Block sizes obtained from vendors' literature and may vary with locality.
7. Average weight of block and cell infill assumed to be 120 pcf.
8. See vendor web sites (on input screen) for more information and specifications.
9. Design height is limited to 12 feet or 18 blocks, whichever is less. Contact vendor for higher designs or special conditions.
10. Seismic design is per *Seismic Design Manual* cited above. Also see *Methodology/Seismic Design* in *User's Manual*.
11. Vendor specifications or project specifications, whichever is more restrictive, to be followed for construction procedures.
12. Add notes and details for proper drainage.
13. See *User's Manual* Design Example #11 for methodology and sample verification calculations.
14. Final design responsibility is with the project Engineer-of-Record.

DESIGN EXAMPLE 12   **Page 1 of 3**

<u>Cantilevered wall with pier foundation</u>

Use wall design Example #1

    for forces applied to pier

Revise footing/pier cap to 30" deep
    x 5.0 ft. wide

Try pier spacing = 6.0 ft.

Vertical load from wall
    = 5688 x 6.0 = 34,128 lbs.

Total lateral load at top of pier from wall

$$= \frac{45\left(10.0 + 2.5 + 1.0\right)^2}{2} \text{ x } 6.0 = 24{,}604^{\#} \text{ / ft}$$

Assume no lateral support at top of pier

Pier $f_c' = 3000$ psi $f_y = 60,000$ psi

Soil bearing at pier tip = 6,000 psf

Use skin friction = 100 psf,

Assume pier fixity at $\frac{1}{6}$ pier depth

Allow passive for pier: 300 pcf.

Load factor for pier concrete design = 1.6

Assume added lateral load at pier top (creep effect) = 500 lbs.

Assume pier diameter effectiveness multiplier for passive = 2.5

Moment applied at top of pier (unfactored) = 24,604 x $\frac{13.5}{3}$ = 110,718 ft-lbs.

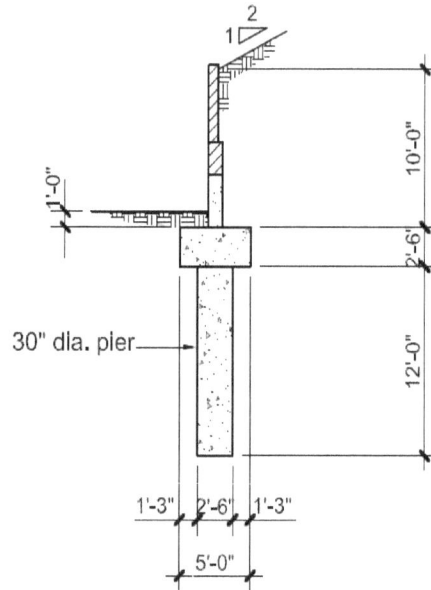

30" dia. pier

1'-3" 2'-6" 1'-3"

5'-0"

10'-0"

2'-6"

12'-0"

1'-0"

DESIGN EXAMPLE 12   **Page 2 of 3**

Try 30" diameter pier spaced 6.0 ft. on center

Determine embedment depth per IBC '15, 1807.3.2.1

$$P = 24,640 \text{ lbs.} \qquad \text{Equiv "h"} = \frac{110,718}{24,604} = 4.50 \text{ ft.}$$

Equation 18.1 requires iteration for "d," therefore assume d = 12.0 ft, then:
   d = 1.5 x 1+ = 1.5 x 10.5 = say 16 ft.

IBC '15 Equation 18-1:

$$d = 0.5A \left[1 + (1 + (4.36 \text{ h/A})^{1/2} \right]$$

$$A = \frac{2.34 \times 24,604}{(300 \times 12) \times 0.33 \times 2.5 \times 2.5} = 7.75$$

$$h = \frac{110.768}{24,604} = 4.5 \text{ ft.}$$

$$d = 0.5 \times 7.75 \left[1 + (1 + 4.36 \times 4.5 / 7.75)^{1/2}\right] = 11.16 \text{ ft.} \quad \textit{Use 12'.0".}$$

   Note: If there is lateral restraint at the top of the pier, then IBC '15, 1807.3.2.2 applies:

$$d = \frac{\sqrt{4.25 M_g}}{S_3 h}$$

   where:

   M_2 = *Moment in the post at grad, in foot-pounds (kN-m).*

   S_3 = *Allowable lateral soil-bearing pressure as set forth in Section 1806.2 based on a depth equal to the depth of embedment in pounds per square foot (kPa).*

Total bearing capacity (neglect pier weight)

$$= \frac{2.5^2 \times 3.14}{4} \times 6000 + 2.5 \times 3.14 \times 100 \times 12 = 39,045 > 34,128.$$

P_V applied at back of footing not applied. (conservative).

Assume stem weight centered over pile.

Pier M_u = [24,604 (4.50 + 12/6)] + 500 (12/6)] 1.6 − [(2.0 x 12.5 x 100 x 6)1.6] = 233,482 ft. − lbs.

DESIGN EXAMPLE 12   **Page 3 of 3**

Check pier ϕM_n per *Whitney Approximation Method

    *ASCE transactions paper 1942 by Charles Whitney

Use 30" diameter with 8 - #8 bars, circular pattern

$$f_c' = 3000 \text{ psi}, \quad f_y = 60,000 \text{ psi}, \quad \phi = 0.90, \quad \text{clearance} = 3"$$

Whitney equivalent rectangular width = (0.80 x 30) = 24 in.

Whitney equivalent "d" = $\frac{2}{3}$ (30) = 20 in.

Whitney equivalent shear area: 24 x 20 = 480 sq. in.

Per ACI 318-14 equations:

$$a = \frac{A_s\, f_y}{.85\, f_c' b} = (8 \times 0.5 \times 0.79 \times 60{,}000) / (0.85 \times 3000 \times 24) = 3.1"$$

$$\phi M_n = 0.90\, A_s\, f_y \left(d - \frac{a}{2} \right) = (0.90 \times 8 \times 0.5 \times .79 \times 60{,}000)$$

$$\left[20 - \left(\frac{3.1}{2} \right) \right] = 3198 \text{ in.-kips} = 266{,}500 \text{ ft-lbs.}$$

Check shear in pier effective area "bd" for 30" circular pier per Whitney approximation:

    $V_u = [(24{,}604 \times 6) + 500]\, 1.6 = 40{,}166 \text{ lbs.}$

    $v_u = 40{,}166 / (24 \times 20) = 83.7 \text{ psi.}$

$$v_{allow} = \phi\, 2 \sqrt{f_c'} = 0.75 \times \sqrt{3{,}000} = 82.16 \cong 83.7 \text{ psi}$$

    $\therefore$ Pier design OK

DESIGN EXAMPLE 12   Report Printout

Basics of Retaining Wall Design

Title   EX-12
Job #
Description...
EX-12

Dsgnr   HB   Date: 2 JUL 2018   Page: 1

This Wall in File: c:\Users\christ\Dropbox\RetainPro 10 Project Files\HB\Hugh.RPX

RetainPro (c) 1987-2018, Build 11.16.36.22
License : KW-06000216
License To : RetainPro Development Center

Cantilevered Retaining Wall   Code: IBC 2015, ACI 318-14, ACI 530-13

Criteria

Retained Height	=	10.00 ft
Wall height above soil	=	0.00 ft
Slope Behind Wall	=	2.00
Height of Soil over Toe	=	12.00 in
Water height over heel	=	0.0 ft

Soil Data

Allow Soil Bearing	=	3,000.0 psf
Equivalent Fluid Pressure Method		
Active Heel Pressure	=	45.0 psf/ft
Passive Pressure	=	350.0 psf/ft
Soil Density, Heel	=	110.00 pcf
Soil Density, Toe	=	110.00 pcf
Footing||Soil Friction	=	0.400
Soil height to ignore for passive pressure	=	12.00 in

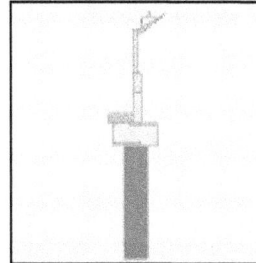

Surcharge Loads

Surcharge Over Heel	=	0.0 psf
Used To Resist Sliding & Overturning		
Surcharge Over Toe	=	0.0 psf
Used for Sliding & Overturning		

Axial Load Applied to Stem

Axial Dead Load	=	0.0 lbs
Axial Live Load	=	0.0 lbs
Axial Load Eccentricity	=	0.0 in

Lateral Load Applied to Stem

Lateral Load	=	0.0 #/ft
...Height to Top	=	0.00 ft
...Height to Bottom	=	1.00 ft
Load Type	=	Wind (W) (Service Level)
Wind on Exposed Stem (Service Level)	=	0.0 psf

Adjacent Footing Load

Adjacent Footing Load	=	0.0 lbs
Footing Width	=	0.00 ft
Eccentricity	=	0.00 in
Wall to Ftg CL Dist	=	0.00 ft
Footing Type		Line Load
Base Above/Below Soil at Back of Wall	=	0.0 ft
Poisson's Ratio	=	0.300

Design Summary

Pier Foundation

Not Restrained at top			
Spacing	=	6.00	ft
Diameter	=	30.00	in
Pier Reinforcing	=	8 - #8	
Embedment	=	12.00	ft
fc	=	3,000	psi
fy	=	60,000	psi
Added Lat. at pier top	=	500.0	lbs
End soil bearing allow	=	6,000	psf
Skin friction	=	100.0	psf
Allow. passive press	=	300	pcf
ignore Pass. Pressure from Pier Top	=	0	ft
Mu	=	235,215	ft-lbs
<phi>Mn	=	266,568	ft-lbs
Vu	=	86.88	psi
<phi>Vn	=	82.16	psi
Total Vertical Load	=	36,566	lbs
Vertical Capacity	=	37,284	lbs
Total Lateral Force to Pier Lateral Restraint	=	25,715.0	

Sliding Stability

Factor of Safety applied to reduce
Allow. Passive Press =   1.0

Vertical component of active lateral soil pressure IS
NOT considered in the calculation of soil bearing

Load Factors

Building Code	IBC 2015, ACI
Dead Load	1.200
Live Load	1.600
Earth, H	1.600
Wind, W	1.000
Seismic, E	1.000

Stem Construction

		3rd	2nd	Bottom
		Stem OK	Stem OK	Stem OK
Design Height Above Ftg	ft =	5.33	3.33	0.00
Wall Material Above "Ht"	=	Masonry	Masonry	Concrete
Design Method	=	ASD	ASD	LRFD
Thickness	=	8.00	12.00	12.00
Rebar Size	=	# 5	# 5	# 7
Rebar Spacing	=	32.00	16.00	16.00
Rebar Placed at	=	Edge	Edge	Edge

Design Data

		3rd	2nd	Bottom
fb/FB + fa/Fa	=	0.511	0.437	0.850
Total Force @ Section				
Service Level	lbs =	490.7	1,001.0	
Strength Level	lbs =	90.0		3,600.0
Moment...Actual				
Service Level	ft-# =	763.9	2,225.6	
Strength Level	ft-# =	270.0		12,000.0
Moment...Allowable	ft-# =	1,494.8	5,093.8	18,288.8
Shear.....Actual				
Service Level	psi =	5.4	7.2	
Strength Level	psi =			31.4
Shear...Allowable	psi =	44.8	45.0	75.0
Anet (Masonry)	in2 =	91.50	139.50	
Rebar Depth 'd'	in =	5.25	9.00	9.56

Masonry Data

		3rd	2nd	Bottom
f'm	psi =	1,500	1,500	
Fs	psi =	32,000	32,000	
Solid Grouting	=	Yes	Yes	
Modular Ratio 'n'	=	21.48	21.48	
Wall Weight	psf =	78.0	124.0	150.0
Short Term Factor	=	1.000	1.000	
Equiv. Solid Thick	in =	7.60	11.60	
Masonry Block Type	=	Medium Weight		
Masonry Design Method	=	ASD		

Concrete Data

		Bottom
Fc	psi =	2,500.0
Fy	psi =	60,000.0

DESIGN EXAMPLE 12 — Report Printout

Basics of Retaining Wall Design

Title   EX-12
Job #
Description
EX-12

Dsgnr   HB

Date   2 JUL 2018

Page   2

This Wall in File: c:\Users\chris\Dropbox\RetainPro 10 Project Files\HB\Hugh.RPX

RetainPro (c) 1987-2018, Build 16.18.06.22
License : KW-06006216
License To : RetainPro Development Center

Cantilevered Retaining Wall   Code: IBC 2015, ACI 318-14, ACI 530-13

Concrete Stem Rebar Area Details

Bottom Stem	Vertical Reinforcing	Horizontal Reinforcing
As (based on applied moment)	0.2885 in2/ft	
(4/3) * As	0.3846 in2/ft	Min Stem T&S Reinf Area 0.959 in2
200bd/fy 200(12)(9.9625)/60000	0.3825 in2/ft	Min Stem T&S Reinf Area per ft of stem Height 0.288 in2/ft
0.0018bh 0.0018(12)(12)	0.2592 in2/ft	Horizontal Reinforcing Options

		One layer of	Two layers of
Required Area	0.3825 in2/ft	#4 @ 9.33 in	#4 @ 18.67 in
Provided Area	0.45 in2/ft	#5 @ 12.92 in	#5 @ 25.83 in
Maximum Area	1.2954 in2/ft	#6 @ 18.33 in	#6 @ 36.67 in

Footing & Pier Data

Toe Width	=	2.00 ft
Heel Width	=	3.00
Total Footing Width	=	5.00
Footing Thickness	=	30.00 in
Pier Diameter	=	30.00 in
Pier Embedment	=	12.00 ft
Footing Toe to C.L. Pier	=	2.50 ft
fc = 2,500 psi Fy = 60,000 psi		
Footing Concrete Density = 150.00 pcf		
Min. As %		0.0018
Cover @ Top 2.00 @ Btm= 3.00 in		

Footing Design Results

		Toe	Heel
Factored Pressure	=	0	0 psf
Mu' Upward	=	0	0 ft-#
Mu' Downward	=	-1,164	5,098 ft-#
Mu: Design	=	-1,164	5,098 ft-#
Actual 1-Way Shear	=	0.39	12.83 psi
Allow 1-Way Shear	=	40.00	40.00 psi
Toe Reinforcing	=	#7 @ 16.00 in	
Heel Reinforcing	=	#6 @ 16.00 in	
Key Reinforcing	=	None Spec'd	

Other Acceptable Sizes & Spacings

Toe   Not req'd: Mu < phi*5*lambda*sqrt(f'c)*Sm
Heel Not req'd: Mu < phi*5*lambda*sqrt(f'c)*Sm
Key

Min footing T&S reinf Area		3.24	in2
Min footing T&S reinf Area per foot		0.65	in2 /ft

If one layer of horizontal bars	If two layers of horizontal bars
#4 @ 3.70 in	#4 @ 7.41 in
#5 @ 5.74 in	#5 @ 11.48 in
#6 @ 8.15 in	#6 @ 16.30 in

Footing Design Results Pier

Footing Torsion, Tu	=	91,804.40 ft-lbs
Footing Allow Torsion, phi Tu	=	61,618.79 ft-lbs

If torsion exceeds allowable, provide supplemental design for footing torsion.

DESIGN EXAMPLE 12   Report Printout

Basics of Retaining Wall Design   Title   EX-12   Page   3
                                               Job #   Dsgnr HB   Date   2 JUL 2018
                                               Description ...
                                               EX-12

This Wall in File: c:\Users\chris\Dropbox\RetainPro 10 Project Files\HB\Hugh.RPX
RetainPro (c) 1987-2018, Build 11.18.06.22
License : KW-06000216   Cantilevered Retaining Wall   Code: IBC 2015, ACI 318-14, ACI 530-13
License To : RetainPro Development Center

Summary of Overturning & Resisting Forces & Moments

Item	....OVERTURNING.... Force lbs	Distance ft	Moment ft-#		....RESISTING.... Force lbs	Distance ft	Moment ft-#
Heel Active Pressure =	4,202.5	4.56	19,144.7	Soil Over Heel =	2,566.7	3.83	9,838.9
Surcharge over Heel =				Sloped Soil Over Heel =	149.7	4.22	632.2
Surcharge Over Toe =				Surcharge Over Heel =			
Adjacent Footing Load =				Adjacent Footing Load =			
Added Lateral Load =				Axial Dead Load on Stem =			
Load @ Stem Above Soil =				* Axial Live Load on Stem =			
=				Soil Over Toe =	220.0	1.00	220.0
				Surcharge Over Toe =			
Total	4,202.5	O.T.M.	19,144.7	Stem Weight(s) =	1,111.8	2.45	2,718.7
				Earth @ Stem Transitions =	171.2	2.83	485.2
=		=		Footing Weight =	1,875.0	2.50	4,687.5
				Key Weight =			
				Vert. Component =			
				Total =	6,094.4 lbs	R.M.=	18,582.4

Pier foundation used. Forces and moments displayed are
applied at top of pier.

* Axial live load NOT included in total displayed, or used for overturning
resistance; but is included for soil pressure calculation.

Vertical component of active lateral soil pressure IS NOT considered in
the calculation of Sliding Resistance.

Vertical component of active lateral soil pressure IS NOT considered in
the calculation of Overturning Resistance.

Tilt

Horizontal Deflection at Top of Wall due to settlement of soil

(Deflection due to wall bending not considered)

Soil Spring Reaction Modulus	250.0 pci
Horizontal Defl @ Top of Wall (approximate only)	0.036 in

The above calculation is not valid if the heel soil bearing pressure exceeds that of the toe,

because the wall would then tend to rotate into the retained soil.

DESIGN EXAMPLE 13 | **Page 1 of 2**

Design Data

Soldier pile, cantilevered

Retained ht. = 10.0 ft.

Backfill slope = 18.4° (3:1)

Soil density = 110 pcf

Soil friction, ϕ = 34°

Surcharge = 100 psf

Pile spacing = 8.0 ft.

Use Rankine soil pressure method

K_a (horiz) = cos 18.4

$$\frac{\text{Cos } 18.4 - \sqrt{cos^2 - cos^2 34}}{\text{Cos } 18.4 + \sqrt{cos^2 - cos^2 34}} = 0.31$$

$K_P = \tan^2\left(45 + \frac{34}{2}\right) = 3.54$

Passive = 3.54 x 110 = 389 psf

$$P_A = \frac{0.31 \times 110 \times 10^2}{2} \times 8.0 = 13{,}640 \text{ lbs. / pier}$$

$P_w = 0.31 \times 100 \times 10 \times 8.0 = 2{,}480$ lbs. / pier

Safety factor to apply to passive = 1.5

Drill hole diameter = 24"

Multiplier for effective diameter = 2.5

$$P_p = P_A = P_w = 13{,}6404 + 2{,}480 = 16{,}120 \text{ lbs.} = \frac{389 \times 0.67 \times 2.0 \times 2.5 \times d^2}{2}$$

$$d = \sqrt{\frac{16{,}120 \times 2}{389 \times 0.67 \times 2.0 \times 2.5}} = 4.97 \text{ ft.}$$

$$M_{max} = P_A\left(\frac{H}{3} + \frac{2d}{3}\right) + P_w\left(\frac{H}{2} + \frac{2d}{3}\right)$$

$$= 13{,}640\ (3.33 + 3.32) + 2480\ (5 + 3.32) = 111{,}340 \text{ ft-lbs. / pier}$$

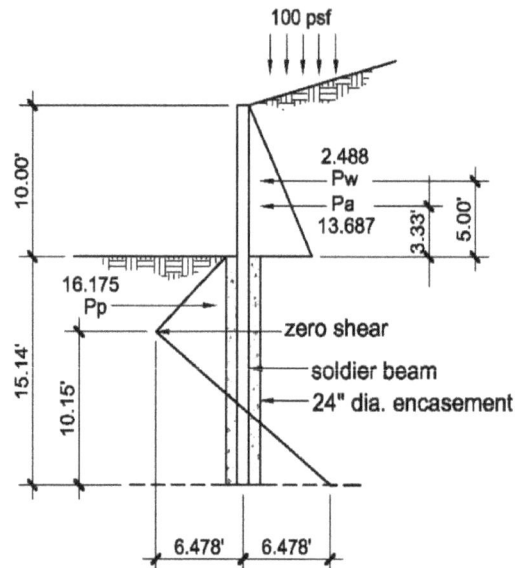

100 psf

2.488 Pw

Pa 13.687

16.175 Pp

zero shear

soldier beam

24" dia. encasement

10.00'

15.14'

10.15'

3.33'

5.00'

6.478' | 6.478'

DESIGN EXAMPLE 13   **Page 2 of 2**

M_{max} = resisted by moment couple = $F_P \dfrac{2}{3}$ x D $D = \dfrac{M_{max}}{0.67\, F_p}$

Mom. Resisting couple = $(0.67D) \left(\dfrac{D}{2} \, x \, 389 \, x \, 0.67 \, x \, 2.0 \, x \, 2.5 \, x \, 4.98 \, x \, 0.5 \right) = D^2 \,(1089)$

$$D \;=\; \sqrt{\dfrac{111{,}637}{1089}} = 10.12$$

Total embedment required = 4.99 + 10.12 = 15.1 ft.

Select steel beam

Per AISC 13th Edition, 2007, LRFD

Factored moment for LRFD = 111,340 x 1.6 = 178,143 ft-lbs. / pier

Assume lateral support (designed to assess)

Select W10 x 49

C_{mn} = 227,000 ft-lbs. > 177,878 ft-lbs.

Check Lagging at, say, 8 ft. depth

Assume $M = \dfrac{WL^2}{10} = \dfrac{[(8 \, x \, 0.31 \, x \, 110) + (0.31 \, x \, 100)]8^2}{10} = 1944$ ft-lbs.

$S_{req'd}$ $= \dfrac{1944 \, x \, 12}{900} = 25.9$ in^3

Use 4 x 12 $\left(S = \dfrac{11.5 \, x \, 3.5^2}{6} = 23.5 \right)$ Consider OK

DESIGN EXAMPLE 13   Report Printout

Basics of Retaining Wall Design

Title   EX-13
Job # :   Dsgnr: HB   Page : 1
Description....   Date: 9 FEB 2018
EX-13

This Wall in File: C:\Users\chris\Dropbox\RetainPro 10 Project Files\Hugh.RPX

RetainPro (c) 1987-2018, Build 11.18.1.19
License : KW-06050001
License To : RETAIN PRO SOFTWARE

Soldier Pile Retaining Wall   Code: IBC 2015,ACI 318-14,ACI 530-13

Summary

Wall height (retained height), ft	10.00
Backfill slope	18.40
Soil Density, pcf	110.00
Soil Phi angle	34.00
Ka (horizontal)	0.31
Surcharge, psf	100
Allow. Passive, psf / ft. depth	369
Apply S.F. to Passive	1.5
Pile Spacing, ft	8.0
Drilled Diameter, in.	24.00
Multiplier to Passive Wedge	2.5
Required Embedment, ft	15.14
Embedment Used, ft	16.00
Moment in Pile Max, ft-lbs	111,739
Max. Mom. Factored 1.6, LRFD	178,783
Soldier Beam Selection	W10x49
Lagging Depth, ft	8.00
Lagging Selection	4x12

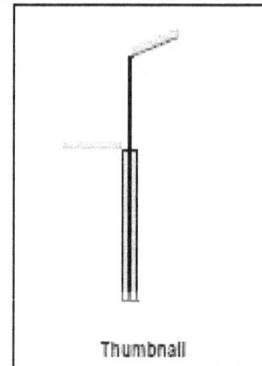

Thumbnail

Selected option to drill hole, insert soldier beam, and encase beam in lean concrete.

DESIGN EXAMPLE 14   **Page 1 of 3**

<u>Gabion Wall</u>

<u>Criteria</u>

Height of each course = 36"

Retained height = 12.0 ft.

Wall tilt from vert. = 6°

Surcharge = 0.0 psf

Density of cages (or blocks) = 120 pcf

Density of backfill = 110 pcf

Backfill slope: Level

Soil friction angle = 34°

Soil/block friction angle assumed zero (Conservative)

Allow. soil bearing = 2500 psf

Coef. of inter-block course friction = 0.70

Coef. of friction at base/soil interface = 0.45

Use Coulomb equation for K_a (horiz)

α angle for Coulomb = $90° - \left[tan^{-1} \left(\dfrac{4.5}{12.0} \right) \right] + 6.0 = 75.4°$

$K_a = 0.39$

K_a (horiz.) = 0.39 [cos (20.6 + 6.0)] = 0.38

<u>Check Course #4 (top)</u>

Resisting moment = $3.0 \times 4.5 \times 120 \times \dfrac{4.5}{2} \times \cos 6° = 3625$ ft.-lbs.

Lateral on course 4 = $0.38 \times 110 (12.0-9.0)^2 \times 0.5 \times \cos 6° = 187$ lbs.

OTM for #4 = 187 x 3.0 x .33 = 187 ft. lbs.

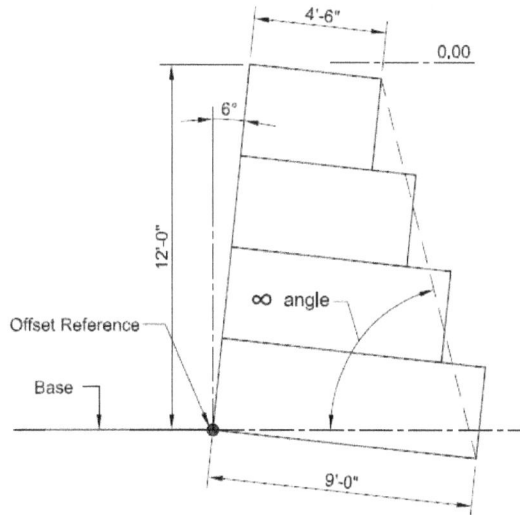

DESIGN EXAMPLE 14 Page 2 of 3

Stability S.F. $= \dfrac{3625}{187} = 19.4$ OK

Sliding resistance $= 0.70 \times 3.0 \times 4.5 \times 120 = 1134$ lbs.

S.F. for sliding $= \dfrac{1134}{187} = 6.06$ 0k

(Repeat this for courses 2 and 3)

Check Stability at Base Course

Distance from reference point to c.g. of courses:

$\#4 = \left(9\,tan\,6° + \dfrac{4.5}{2}\right)cos\,6° = 3.17$ ft.

$\#3 = \left(6\,tan\,6° + \dfrac{6}{2}\right)cos\,6° = 3.59$ ft.

$\#2 = \left(3\,tan\,6° + \dfrac{7.5}{2}\right)cos\,6° = 4.02$ ft.

$\#1 =$

Overall Resisting Moment

$= 3.17\,(3 \times 4.5 \times 120) + 3.59\,(3 \times 6 \times 120)$

$+ 4.02\,(3 \times 7.5 \times 120) + 4.5\,(3.0 \times 9.0 \times 120) = 38,324$ ft-lbs.

Overturning Moment

$= \left(0.38\,x\,110\,x\,12^2\,x\,0.5\,x\,\dfrac{12}{3}\right)cos^2\,6° = 11,918$ ft-lbs.

Stability S.F. $= \dfrac{38,324}{11,918} = 3.22$

Sliding Resistance

Total vertical load $= 4.5 \times 3 \times 120 + 6.0 \times 3 \times 120 + 7.5 \times 3 \times 120$

$+ 9.0 \times 3 \times 120 = 9,720$ lbs.

Resistance $= 9,720 \times 0.45 = 4374$ lbs.

Lateral force $= (0.38 \times 110 \times 12^2 \times 0.5)\,cos\,6° = 2980$ lbs.

Sliding S.F. $= \dfrac{4374}{2980}\,1.47$

DESIGN EXAMPLE 14   **Page 3 of 3**

Soil Bearing

Total vertical load = 9,720 lbs.

Dist. To c. g. vertical load

$$= \frac{38,324 - 11,918}{9720} = 2.72 \text{ ft.}$$

$$e = \frac{9}{2} - 2.72 = 1.78 > \frac{9}{6} = 1.5$$

$$\text{Soil bearing} = \frac{9720}{.75 \times 9.0 - 1.5 \times 1.78} = 2362 \text{ psf.} - \text{N/A e} > \frac{B}{6}$$

DESIGN EXAMPLE 14   **Report Printout**

Title   : EX-14   Page: _____
Job #   : 914   Dsgnr: Date: MAY 29, 2011
Description:

This Wall In File: c:\users\brooks\documents\retainpro9\ex

Gabion Retaining Wall Design   Code: IBC 2009

Summary

Input Values:

Course Height (Gabion/Block), in:	36.00
Retained Height, ft:	12.00
Wall Tilt from Vertical, degrees:	6
Surcharge, psf:	0
Density, Gabion Infill or Block, pcf:	120.00
Density of Backfill, pcf:	110.00
Backfill Slope, Degrees:	0.00
Soil Friction Angle, Phi:	34.00
Allow Soil Bearing, psf:	2,500
Coef. of Friction on Soil:	0.45
Coef. of Interblock Friction:	0.70

Result Values

Ka (horiz):	0.38
Coulamb Alpha Angle:	69.44
Act. Soil Bearing Pressure, psf:	2,379
Lateral Force, Earth, lbs:	3,020
Lateral Force, Surcharge, lbs:	0
Total Lateral Force, lbs:	3,020
Resisting Moment, ft-lbs:	38,385
Overturning Moment, ft-lbs:	11,893
Overturning Ratio:	3.23
Sliding Ratio:	1.46

Thumbnail

Course	Height	Offset	Length	Vertical	Dist	RM	Lateral	OTM	Stab. S.F.	Sliding S.F.
4	9.00	0.00	4.50	1620.0	3.18	3625	187.7	186	19.49	6.04
3	6.00	0.00	6.00	3780.0	3.61	10578	750.8	1487	7.11	3.52
2	3.00	0.00	7.50	6480.0	4.04	21832	1689.4	5017	4.35	2.68
1	0.00	0.00	9.00	9720.0	4.48	38365	3003.3	11893	3.23	1.46

Column Title Descriptions:

Course:	These are numbered in ascending order and cannot exceed 10.
Height:	Measured from bottom of first (base) course.
Offset:	Measured from front edge bottom course.
Length:	Of cages or blocks in course.
Vertical:	Accumulated load from courses above.
Dist.:	From front edge bottom course to c.g. of course.
RM:	Resisting moment at level selected.
Lateral:	Accumulated lateral from earth pressure and surcharge.
OTM:	Accumulated overturning moment above Height.
Stab. S.F.:	RMl / OTM
Sliding S.F.:	Lateral / Vertical * (Coef. Interblock friction).

DESIGN EXAMPLE 14

Report Printout

Title   · EX-14
Joh #   , 914   Dsgnr:
Description....

Page: _____
Date: MAY 29,2011

This Wall in File: c:\users\brooks\documents\retainpro9\ex

Gabion Retaining Wall Design

Code: IBC 2009

Notes:

1. All courses are the same height and infill density.
2. If wall depth is uniform consider using segmental wall module, gravity, select "custom".
3. Concrete blocks may be used in lieu of Gabion cages.
4. Offset of successive layers limited to one-half course height. Earth side face flush.
5. Coulomb equation used for active pressure. Wall friction angle assumed zero.
6. If wall is tilted (battered), model effect by successive offsets (tan beta times course height).
7. This design not valid for reinforced soils (MSE). Use SRW module.
8. Vendor specifications may apply.

APPENDIX

Appendix A - Summary of Design Equations with Code References

(Code referenced below: IBC 2018, ACI 318-14, TMS 402/602-16)

Concrete (SD)

ϕ = .90 for flexure [ACI – 21.2.2]

$\quad$ = .75 for shear and torsion [ACI – 21.2.1]

β = 0.85 for $f_c' > 4,000$ [ACI – 22.2.2.4.3]

$\epsilon_t \geq$ 0.004 [ACI 9.3.3.1]

ρ_{min} = $200/fy$ [ACI – 9.6.1.2]

E_s = 29,000,000 psi [ACI – 20.2.2.2]

E_c = 57,000 $\sqrt{f_c}$ [ACI – 19.2.2.1]

n = A_s/E_c

a = $\dfrac{A_z f_y}{.85 f_c b}$ [ACI – 22.2.2]

The general solution for A_s [CRS I Handbook, 14 –41]

$$A_s = \frac{1.7 f_c\, bd}{2 f_y} - \frac{1}{2}\sqrt{\frac{2.89\,(f_c\,bd)^2}{f_y^2} - \frac{6.8\,f_c\,b\,M_n}{\phi f_y^2}}$$

M_n = in - kips

For b = 12", f_y = 60 ksi, this reduces to:

$$A_s = 0.17\, f_c d - \sqrt{0.29\,(f_c')^2 - .0063\, f_c\, M_n}$$

Moment capacity

$$M_n = A_s f_y \left(d - \frac{a}{2}\right)$$

a = $\dfrac{A_s}{bd}$

ϕ = 0.90

A_s = Area per foot of steel of wall

f_y = 40,000, 60,000 or 80,000 psi

d = effective depth to reinforcing

Shear Capacity

v_c = $2\phi\sqrt{f_c'}$ [ACI – 22.5.5.1]

ϕ = 0.75

$$v_u = \frac{V_u}{bd}$$

Applied Moment

$M_u \leq \phi M_n$

Hook bar embedment

$$l_{dh} = \frac{0.02\, f_y\, d_b \times 0.7}{\sqrt{f_c'}}\left(\frac{A_s\ req\ d}{A_s\ provided}\right)\ [ACI – 25.4.3.1]$$

$\quad$ or minimum of .8d_b or 6"

Development length

l_d (#6 and smaller)

$$= \frac{0.024\, d_b\, f_y}{\sqrt{f_c'}}\left(\frac{A_s\ req'd}{A_s\ provided}\right)\ [ACI – 25.5.2.1]$$

l_d (#7 and larger)

$$= \frac{0.03\, d_b\, f_y}{\sqrt{f_c'}}\left(\frac{A_s\ req'd}{A_s\ provided}\right)[ACI – 25.5.2.2]$$

Lap length Class B splice 1.3 l_d [ACI –25.4.2.1]

Allowable stress in plain concrete

Plain concr tension = $5\sqrt{f_c'}$ [ACI – 14.5.2.1]

Plain concr shear = $1.33\sqrt{f_c'}$ [ACI – 14.5.5]

Plain concr = ϕ flexure/shear = 0.6 [ACI – 21.2.1]

$\phi M_n \geq M_u$ [ACI – 14.5.1.1]

Masonry (ASD)

E_s = 29,000,000 [TMS – 4.2.2]

E_m = 900 f_m' [TMS – 4.2.2]

n = E_s/E_m

F_b = .45 f_m' [TMS – 8.3.4.2.2]

v_u = [Refer to TMS – 8.3.5.1.3]

f_s = 20,000 psi Grade 40

or 32,000 psi Grade 60 [TMS – 8.3.3.1]

$p = A_s/bd$

$k = \sqrt{(np)^2 + 2np} - np$

$j = 1 - k/3$

$M_s = F_s A_s j d$

$M_m = F_b bd^2 \left(\dfrac{kj}{2}\right) = Kbd^2$, where $K = \dfrac{kj}{2}$

$f_v = V/bd$ [TMS – 8.3.5]

F_v = [Refer to TMS – 9.3.4.1.2.1]

$l_d = .002 d_b F_s$ (but not less than 12")
[IBC – 2107.2.1]

Masonry (LRFD)

$a = \dfrac{A_s f_y}{0.80 f_m b}$ [TMS – 9.3.2]

$M_n = A_s f_y \left(d - a/2\right)$

ϕ flexure 0.90 [TMS – 9.1.4.4]

ϕ shear 0.80 [TMS – 9.1.4.5]

$V_n = 4 A_n \sqrt{f_m'}$ [TMS – 9.3.4.1.2]

A_n = Effective section area

$l_d = \dfrac{0.13 d_b^2 f_y}{K \sqrt{f_m'}}$ [TMS – 6.1.5.1.1]

K = Masonry cover

For #6 to #7 bars multiply by 1.3 [TMS – 6.1.5.1.1]

Rankine equation for active pressure
[Bowles pg. 603]

$$K_a = \cos\beta \; \frac{\cos\beta - \sqrt{\cos^2\beta - \cos^2\phi}}{\cos\beta + \sqrt{\cos^2\beta - \cos^2\phi}}$$

K_a (horiz.) $= \cos\beta \; K_a$

β = Angle of backfill slope

ϕ = Angle of internal friction

Coulomb equation for active pressure [Bowles pg. 595]

$$K_a = \frac{\sin^2(\alpha+\phi)}{\sin^2\alpha \sin(\alpha-\delta)\left[1+\sqrt{\dfrac{\sin(\phi+\delta)\sin(\phi-\beta)}{\sin(\alpha-\delta)\sin(\alpha+\beta)}}\right]^2}$$

K_a (horiz.) $= \cos\delta \; K_a$ if $\alpha = 90°$ If $\alpha \neq 90°$, then

K_a (horiz) $= \cos(90 \pm \alpha - \delta)$

β = Angle of backfill slope
ϕ = Angle of internal friction of the soil
α = Wall slope angle from horizontal (90° for vertical face, α = < 90° if the back of wall is battered outward or > 90° if wall battered inward)
δ = Angle of friction between soil and wall (Usually assumed to be $2/3\phi$ to $1/2\phi$)

Mononobe-Okabe equation [Bowles pg 643]

K_{AE} = active earth pressure coefficient, static + seismic

$$= \frac{\sin^2(\alpha+\theta-\phi)}{\cos\theta \sin^2\alpha \sin(\alpha+\theta+\delta)\left[1+\sqrt{\dfrac{\sin(\phi+\delta)\sin(\phi-\theta-\beta)}{\sin(\alpha+\delta+\theta)\sin(\alpha-\beta)}}\right]^2}$$

Where $\theta = \tan^{-1} k_h$, α = wall slope to horiz. (90° for a vertical face), ϕ = angle of internal friction, β = backfill slope, and δ = wall friction angle.

The horizontal component of $K_{AE} = \cos K_{AE}$

FOR NOMENCLATURE SEE APPENDIX I

Appendix B

UNIFIED SOIL CLASSIFICATION SYSTEM – (USCS)

MAJOR DIVISION			GROUP SYMBOL	LETTER SYMBOL	GROUP NAME
COARSE GRAINED SOILS CONTAINS MORE THAN 50% FINES	GRAVEL AND GRAVELLY SOILS MORE THAN 50% OF COARSE FRACTION RETAINED ON NO. 4 SIEVE	GRAVEL WITH <5% FINES		GW	Well-graded GRAVEL
				GP	Poorly graded GRAVEL
		GRAVEL WITH BETWEEN 5% AND 15% FINES		GW-GM	Well-graded GRAVEL with silt
				GW-GC	Well-graded GRAVEL with clay
				GP-GM	Poorly graded GRAVEL with silt
				GP-GC	Poorly graded GRAVEL with clay
		GRAVEL WITH ≥ 15% FINES		GM	Silty GRAVEL
				GC	Clayey GRAVEL
	SAND AND SANDY SOILS MORE THAN 50% OF COARSE FRACTION PASSING ON NO. 4 SIEVE	SAND WITH <5% FINES		SW	Well-graded SAND
				SP	Poorly graded SAND
		SAND WITH BETWEEN 5% AND 15% FINES		SW-SM	Well-graded SAND with silt
				SW-SC	Well-graded SAND with clay
				SP-SM	Poorly graded SAND with silt
				SP-SC	Poorly graded SAND with clay
		SAND WITH ≥ 15% FINES		SM	Silty SAND
				SC	Clayey SAND
FINE GRAINED SOILS CONTAINS MORE THAN 50% FINES	SILT AND CLAY	LIQUID LIMIT LESS THAN 50		ML	Inorganic SILT with low plasticity
				CL	Lean inorganic CLAY with low plasticity
				OL	Organic SILT with low plasticity
		LIQUID LIMIT GREATER THAN 50		MH	Elastic inorganic SILT with moderate to high plasticity
				CH	Fat inorganic CLAY with moderate to high plasticity
				OH	Organic SILT or CLAY with moderate to high plasticity
HIGHLY ORGANIC SOILS				PT	PEAT soils with high organic contents

NOTES:
2) Based upon ASTM D2488

3) Solid lines between soil descriptions indicate change in interpreted geologic unit. Dashed lines indicated stratigraphic change within the unit.

3) Fines are material passing the U.S.Std. #200 Sieves.

First Letter	Definition
G	Gravel
S	Sand
M	Silt
C	Clay
O	Organic

2nd Letter	Definition
P	Poorly graded u(uniform particle sizes)
W	Well-graded (diversified particle sizes)
H	High plasticity
L	Low plasticity

Appendix C - Masonry Design Data

Rebar Position Depth for Masonry, Default Values.

Thickness	Rebar Depth (in)	
	Center	Edge
6"	2.75"	2.75"
8"	3.75"	5.25"
10"	4.75"	7.25"
12"	5.75"	9.0"
14"	6.75"	11.0"
16"	7.75"	13.0"

Masonry Equivalent Solid Thickness (inches)

Thickness (inches)	Grout Spacing					
	8"	16"	24"	32"	40"	48"
6	5.6	4.5	4.1	3.9	3.8	3.7
8	7.6	5.8	5.2	4.9	4.7	4.6
10	9.6	7.2	6.3	5.9	5.7	5.5
12	11.6	8.5	7.5	7.0	6.7	6.5
14	13.6	9.9	8.7	8.1	7.6	7.4
16	15.6	11.6	10.1	9.5	8.6	8.3

Wall Thickness		Concrete Masonry Units											
Solid Grouted Wall		Lightweight 103 pcf				Medium Weight 115 pcf				Normal Weight 135 pcf			
		6"	8"	10"	12"	6"	8"	10"	12"	6"	8"	10"	12"
		52	75	93	118	58	78	98	124	63	84	104	133
Vertical Cored Grouted at:	16" o.c.	41	60	69	88	47	63	80	94	52	66	86	103
	24" o.c.	37	55	61	79	43	58	72	85	48	61	78	94
	32" o.c.	36	52	57	74	42	55	68	80	47	58	74	89
	40" o.c.	35	50	55	71	41	53	66	77	46	56	72	86
	48" o.c.	34	49	53	69	40	45	64	75	45	55	70	83

Appendix D - Development and Lap Lengths

Lap Splice Lengths[1] and Hooked Bar Embedments (inches)

Bar Size		Masonry [2] $f'_m = 1500$ psi		Concrete [5]		
		$f_s = < 0.8\,F_S$ [2]	$f_s => 0.8\,F_s < F_S$ [3]	2000 psi	3000 psi	4000 psi
#4	L	32	48	34.9	28.5	24.7
	H[4]			13.4	11.0	9.5
#5	L	40	60	43.6	35.6	30.8
	H[4]			16.8	13.7	11.9
#6	L	48	72	52.3	42.7	37.0
	H[4]			20.1	16.4	14.2
#7	L	56	84	76.3	62.3	54.0
	H[4]			23.5	19.2	16.6
#8	L	64	96	87.2	71.2	61.7
	H[4]			26.8	21.9	19.0

Source: Reinforced Masonry engineering Handbook; Amrhein, 5th edition.

(1) F_s for Grade 60 = 32,000 psi per TMS 602-16.
(2) Lap length per IBC '18, 2107.2.1 = 0.002 $d_b f_s$
(3) For the higher f_s increase by 50% per IBC '15, 2107.2.1
(4) Assume 90° bend at bottom extending 12 inches.
(5) Per ACI 318-14, 25.4.2.2

Appendix E - Sample Construction Notes

Brief specifications, or notes, should accompany any retaining wall design. A checklist for items to include:

> Reference to foundation investigation report recommendations (if applicable)
> Excavating / grading requirements
> Concrete strength
> Masonry
> Mortar
> Grout
> Reinforcing, including placement requirements
> Soil bearing value and special requirements
> Inspections
> Drainage

And here are a few additional notes that will help solve problems and keep you out of trouble:

1. Should a discrepancy arise between the drawings and field conditions, or where a detail is doubtful of interpretation or an unanticipated field condition be encountered, the structural engineer shall be called right away for procedure to be followed which shall be confirmed in writing by the structural engineer with copies to all parties.

2. Wherever there is a conflict between details and specifications, or between details, or where doubtful of interpretation, the most restrictive shall govern, as determined by the structural engineer.

3. The contactor and each subcontractor shall visit the site and consider field conditions affecting the work depicted on the plans, and his submission of a bid indicates his acceptance of such conditions.

4. The contractor shall assure that each subcontractor has copies of latest plan revisions and is kept current with any change orders or directives affecting the subcontractors work.

And your experience will add more!

Appendix F. - Conversion Factors

English – S.I. – Metric Conversions		
Multiply	by	to get
inches	2.54	cm (centimeters)
feet	0.305	m (meters)
centimeters	0.394	Inches
centimeters	10	mm (millimeters)
meters	3.28	feet
psf	47.9	kPa (pascals)
psi	6.89	kPa (kilopascals)
pcf	16.0	kg/m^2 (kilograms per cubic meter)
psf/ft	0.157	kPa/m (kilopascals per meter)
in-lbs	0.113	Nm (newton meters)
ft-lbs	1.36	Nm (newton meters)
pounds	4.45	N (newtons)
kip	4.45	kN (Kilo Newtons)
lbs per lin ft	1.49	kg/m (kilograms per meter)
inches	25.4	mm (millimeters)
milimeters	.039	inches
Pascals	1.0	N/m^2 (newtons per square meter)
$Newtons/m^2$	1.0	Pa (pascals)

Common Equivalents	
English	S.I.
1,500 psi	= 10.34 mPa
2,000 psi	= 13.8 mPa
2,500,psi	= 17.24 mPa
3,000 psi	= 20.7 mPa
24,000 psi	= 165 mPa
60,000 psi	= 414 mPa
100 psf	= 4.788 mPa
1,000 psf	= 47.9 mPa

Appendix G - Reinforcing Bar Basics and US/Metric Conversions

In the United States reinforcing bars are manufactured in size designation #4 through #11, where the number indicates the number of eight inches in the diameter. A #5 bar means the diameter is 5/8 inch. The bars are *deformed* with "speed bumps" along their length to increase bond to the concrete. The US system is also known as the Imperial system from its English derivation.

Note: Although the term *reinforcing steel* is used throughout this test, the commonly used term is its contraction to "*rebar.*"

Outside the US the *metric* system is commonly used and is also known as the *S.I. System* (from its French roots that established *International System of Units*). Metric bar sizes use a different number system and termed "soft metric" meaning it is not an exact conversion diameter but "rounded" to standardize manufacturing. See table below. "Hard Metric" uses exact diameters and areas but is rarely used due to non-standard manufacturing.

Reinforcing Bar Conversions ("soft" metric)			
U.S. Bar No.	Metric Bar No.	Diameter in/mm	Area in^2/mm^2
3	10	0.375/9.5	0.11/71
4	13	0.50/12.7	0.20/129
5	16	0.625/15.9	0.31/199
6	19	0.75/19.1	0.44/284
7	22	0.875/22.2	0.60/387
8	25	1.00/25.4	0.79/510
9	29	1.125/28.7	1.00/645
10	32	1.25/32.3	1.27/819
11	36	1.375/38.8	1.56/1006

Grade designation per ASTM A615 showing minimum yield strengths in both US and Metric are shown below. There also appears to be a growing use of 80 ksi and 90 ksi reinforcing – and even 100 ksi – which could offer economics.

Inch-pound Grade	Metric grade	Minimum Yield Strength	
		In pounds Per square inch	In Megapascale
Grade 40	Grade 280	40,000	280
Grade 60	Grade 420	60,000	420
Grade 75	Grade 520	75,000	520

For more information visit *Concrete Reinforcing Steel Institute* at **www.crsi.org**. Their *Manual of Standard Practice* is an industry standard.

Appendix H. - Reference Bibliography

1. International Building Code, 2018, International Code Council, Inc., Falls Church, VA.

2. Building Code Requirements for Masonry Structures (TMS 402/602-16), published jointly by the American Concrete institute and the American Society of Civil Engineers.

3. ACI 318- 14, published by the American Concrete Institute.

4. NEHRP Recommended Provisions for Seismic Regulations for New Buildings and Other Structures, Part 2 – Commentary. 2015 Edition. FEMA.

5. Minimum Design Loads for Buildings and Other Structures, ASCE 7-16.

6. Foundation Analysis and Design, Fifth Edition, by Joseph E. Bowles, published by McGraw-Hill.

7. Reinforced Masonry Engineering Handbook, Fifth Edition, by J. Amrhein, published by the Masonry Institute of America.

8. CRSI Handbook, 2012, published by Concrete Reinforcing Steel Institute.

9. Reinforced Concrete Design, Sixth Edition, Wang & Salmon, published by Harper & Row.

10. Principles of Foundation Engineering, 5th Edition, Braja Das, PWS-KEWT.

11. Introductory Soil Mechanics and Foundations: Engineering, 4th Edition, Sowers, Macmillan, 1979.

12. Foundations and Earth Structures, NAVFAC Design Manual 7.02, 1986.

13. Foundation Engineering, 2nd Edition, Peck, Hansen, Thornburn, Wiley, 1974.

14. Soil Mechanics in Engineering Practice, Tarzaghi and Peck, Wiley, 1967.

15. Foundation Engineering Handbook, Winterkorn & Fang, Van Nostrand Reinhold Company, 1975.

16. Construction Guide for Soil and Foundations, 2nd. Edition, Ahlvin and Smoots, Wiley, 1988.

17. Soil and Foundations for Architects and Engineers, Duncan, Van Nostrand Reinhold, 1992.

18. Foundation Design, Teng, Prentice Hall, 1962.

19. Soil Mechanics Technology, Truitt, Prentice Hall, 1983.

20. Design and Performance of Earth Retaining Structures, ASCE Paper by Robert Whitman, 1990.

21. Lateral Stresses & Design of Earth-Retaining Structures, ASCE Paper, Seed and Whitman, 1970.

22. Seismically Induced Lateral Earth Pressures on a Cantilevered Retaining Wall, Green et al, 2003, Sixth US Conference on Earthquake Engineering.

23. Seismic Analysis of Cantilever Retaining Walls, Phase I, Michigan University Dept. of Civil Engineering, 2002, National Technical Information Service.

24. California Trenching & Shoring Manual, California Dept. of Transportation, 2011.

25. Segmental Retaining Walls, 3rd Edition, National Concrete Masonry Assn. (NCMA).

26. Segmental Retaining Walls Seismic Design Manual, 2010, NCMA.

Appendix I: Notations & Symbols

a	=	depth of equivalent rectangular stress block for Strength Design.
AASHTO	=	American Association of State Highway & Transportation Officials
A_s	=	effective cross-section area of reinforcement in a column or flexural member, square inches.
ACI	=	American Concrete Institute.
ASCE	=	American Society of Civil Engineers.
ASD	=	Allowable Stress Design
ASTM	=	American Society for Testing and Materials.
b	=	width of rectangular member.
B	=	width of footing.
c	=	coefficient that determines the distance to the neutral axis in a beam in strength design.
cm	=	centimeter
CMU	=	Concrete masonry unit.
d	=	depth of reinforcing from compression edge.
DL	=	dead load.
e	=	eccentricity measured from the vertical axis of a section to the resultant force.
E_m	=	modulus of elasticity of masonry in compression, psi.
E_s	=	modulus of elasticity of steel = 29,000,000 psi.
FEMA	=	Federal Emergency Management Agency.
f_a	=	actual axial compressive stress due to axial load psi.
f_b	=	actual flexural stress in masonry extreme fiber due to bending moment, psi.
f_c	=	actual compression strength in concrete or grout

f'_c	=	specified compressive stress in concrete in psi.
f_m	=	actual compressive strength of masonry
f'_m	=	specified compressive strength of masonry, psi.
f_r	=	modulus of rupture, psi.
f_s	=	computed stress in reinforcement due to design loads, psi.
f'_s	=	stress in compressive reinforcement in flexural members, psi.
f_y	=	yield strength of reinforcement, psi.
f_t	=	flexural tensile stress in masonry, psi.
f_v	=	actual shear stress, psi.
F_a	=	allowable axial compressive stress, psi.
F_b	=	allowable flexural compressive stress, psi.
F_s	=	allowable stress in reinforcing.
F_v	=	allowable shear stress, psi
h	=	effective height of wall
IBC	=	International Building Code.
ICC	=	International Code Council
I_g,	=	gross section moment of inertia, in^4.
I_{cr}	=	moment of inertia of cracked Chapter, in^4.
k_h	=	horizontal seismic acceleration factor.
K_a	=	active earth pressure cueficient, static
K_{ae}	=	active earth pressure, static and seismic.
K_p	=	passive earth pressure coefficient
Kip	=	1000 pounds.
kN	=	kilonewtons.

K_o	=	At-rest earth pressure coefficient.	plf	=	pounds per linear foot.
K_p	=	passive earth pressure coefficient.	psi	=	pounds per square inch.
kPa	=	kilopascals	psf	=	pounds per square foot.
ksi	=	Kips per square inch	RM	=	Resisting Moment
ksf	=	Kips per square foot	r	=	radius of gyration.
Kg	=	kilogram.	S	=	section modulus.
l_d	=	required development length of the reinforcement, inches.	S_D	=	strength design.
			SF	=	Safety Factor
l_{hd}	=	hooked bar development length, inches	SI	=	International System of Measurement (Metric).
LL	=	live load.	SRW	=	Segmental Retaining Wall
LRFD	=	Load Resistance Factor Design.	USGS	=	United States Geological Society
MIA	=	Masonery Institute of America	TMS	=	The Masonry Society
MSJC	=	Masonary Standards Joint Committee	v	=	shear stress, psi
MSE	=	Mechanically Stabilized Earth.	v_u	=	factored shear stress, psi.
M_m	=	the moment of the compressive force in the masonry about the centroid of the tensile force in the reinforcement.	v_n	=	allowable shear stress psi.
			V	=	total design shear force or vertical, lbs.
			V_u	=	factored shear force at section.
M_n	=	nominal moment strength of a section.	w	=	uniformly distributed load.
M_{or}	=	overturning moment.	W_p	=	total vertical weight or component thereof.
MPa	=	megapascals.	α	=	In Coulomb equation, clockwise angle from horiztonal to back face of wall (90^0 if wall is vertical).
M_R	=	resisting moment.			
M_s	=	the moment of the tensile force in the reinforcement about the centroid of the compressive force in masonry.	β	=	angle of the backfill slope from a horizontal level plane.
			γ	=	unit weight of soil, pcf.
M_u	=	factored actual moment.	δ	=	angle of the wall friction to a horizontal level plane.
N	=	Newtons.			
NCMA	=	National Concrete Masonry Association.	△	=	deflection of element.
			ρ	=	reinforcing ratio
NEHRP	=	National Earthquake Hazard Reduction Program.	μ	=	coefficient of sliding friction.
OTM	=	Overturning Moment.	φ	=	angle of internal friction of soil degrees. Also strength reduction factor
PGA	=	Peak Ground Acceleration			
pcf	=	pounds per square foot per foot or pounds per cubic foot.	ω	=	batter angle of SRW from vertical.

NOTE: SOME SYMBOLS MAY HAVE DIFFERENT MEANINGS IN DIFFERENT CONTEXTS.

Appendix J: Glossary

Abutment

A relatively massive structural element usually of concrete at the end of a bridge or arch. Its function is to support or otherwise restrain the end of the bridge or arch against movement.

Adobe

A general term applied to various clay soils of medium to high plasticity which occurred throughout the country particularly in the south and west. May range in consistency from gumbo to very hard.

Alluvium

A general term for a soil which at one time was deposited in river beds, flood plains, etc. In addition to soil the deposits may also contain gravel, cobbles and boulders.

Anchored (tieback) walls

Anchors or tiebacks are often used for higher walls where a cantilevered wall may not be economical. Restraint is achieved by drilling holes and grouting inclined steel rod anchors into the zone of earth behind the wall beyond the theoretical failure plane in the backfill.

Angle of internal friction:

This is the most important value for determining lateral pressure and bearing capacity of granular and cohesive soil. It is a measure of the shearing resistance of the soil because of intergranular friction, obtained from one of several laboratory tests.

Angle of repose

The steepest angle measured from the horizontal that a granular material can be piled without sliding. It varies with the type of material for dry sand for example is approximately $34°$ but can vary with the angularity of the grains.

Arching

The transfer of pressure from a yielding mass of soil onto an adjacent stationary mass of soil.

Auger

A screw-like excavation implement which may be drilled into the ground for recovering soil samples or for making cylindrical holes such as for piles.

Backfill

Earth replaced after having been excavated such as to refill utility trenches or behind retaining walls.

Ballast

In railroad work, the gravel or crushed rock between and underneath the ties.

Base

A general term meaning an underlying support for something such as gravel, or soil under a retaining wall foundation.

Batter Pile

See *Piles.*

Bearing capacity (of soil)

A normal pressure, usually expressed in pounds per square foot or tons per square foot which a given soil strata can support without excessive settlement without an adverse effect on the structure being supported. The term *soil-bearing value* is synonymous but the term *soil pressure* is not. The latter states that actual rather than the permissible bearing pressure under a foundation.

Bearing Pile	See *Piles*.
Bedrock	In-situ, undisturbed earth materials lying below the surface soil which may be firm, loose unconsolidated earth. If the bedrock is shattered or otherwise distorted or broken by tectonic activity can still be classified as bedrock.
Boring Log	In soil testing, a sequential description indicating soil classification, water content, color, etc. of soil retrieve from within a *soil boring*.
Boulders	Individual rocks or rock particles which by general agreement are over 12" in diameter.
Build Pier	A pile like concrete foundation made by drilling a hole into the earth and filling it with concrete. Its purpose is to penetrate poor soil to provide firm bearing on deeper strata. If the bottom of bore is enlarged (belled), it is also known as a caisson, particularly in the west.
Bulking (of soil)	The increase in volume of soil after it has been excavated, caused by the loosening of soil particles.
Buttress retaining walls	These are similar to *counterfort walls*, but the wings project from the outside face of the wall.
Calcium carbonate	The chief constituent of limestone and marble and having the mineral name calcite *limestone and marble*
Calcium Chloride	A white, lumpy, absorbent material chiefly characterized by its ability to assimilate water. It is used on roadways to absorb dust and in concrete to accelerate setting time and to prevent freezing.
Caliche	A type of soil usually associated with a desert environment in which the individual grains are naturally cemented together by calcium carbonate.
California Bearing Ratio	In *soil mechanics* is a test for determining the supporting capacity of the soil under a pavement. A measurement is made of the penetration of the cylindrical plunger into the subgrade material and compared to the resistance to penetration of the same size plunger into a standardized sample.
Capillary Action	The tendency of water, or other liquid, to rise in a small tube or network of spaces such as in a soil caused by the surface tension of the liquid.
Caveat	A notice to take heed such as a clause in a document which is meant to be a warning.
Center of Gravity	A single imaginary point about which an object or shape would "balance" if freely suspended.

Clay	A very fine grained (under 0.001 to 0.005 mm in diameter) earth material (the particles are likely, plate shaped) which is powdery when dry but plastic and moldable when moist. When molded and burned in a kiln it becomes hard and weather resistant. Used for a variety of building materials such as bricks, tiles and pipe. Generally the clay is hydrous aluminum silicade but clays are often both chemically and mineralogically altered, which accounts for the differences in the physical properties of clay.
Clay Soil	A soil in which clay-size particles are the principal constituent, usually 35 percent or more. Clay soil expands when wet and shrinks when drying but properties can vary widely with locality.
Cobbles	A classification of rocks or rock particles ranging in diameter from 3 to 12 inches as opposed to boulders (larger) and gravel (smaller).
Coefficient of Friction	A measure of the resistance of sliding of one material resting on another.
Cohesion	In soil mechanics, the attraction clay particles have to each other. The binding forces are caused by molecular attraction and the effect of overburden stresses in excess of the existing overburden.
Cohesive soil:	Soil classification that derives its strength primarily from the cohesive bond between particles. Examples include fine-grained silts and clays.
Compaction	Artificial densification of soil. By placing fill soil in layers, controlling its water content and compacting it by means such as rollers a soil can be made safe to support a building.
Compaction Test	A test for determining the degree of compaction of filled soil. Formerly, the most commonly used field test was a sand displacement method whereby the soil removed from a small test hole was weighed, its volume determined and its density then compared to a laboratory value to establish relative compaction for that particular soil. Today, in-place measurements are made with nuclear gauges.
Contour	Lines on a topographic map or grading plan each of which represents the same elevation in relation to a reference point.
Core Sample	A sample of rock obtained from a core drill.
Counterfort retaining walls	Counterfort cantilevered retaining walls incorporate wing walls projecting upward from the heel of the footing into the stem. The thickness of the stem between counterforts is thinner (than for cantilevered walls) and spans horizontally, as a beam, between the counterfort (wing) walls.
Creep (of soil)	A very slow, downhill movement of soil caused by a combination of gravitational and weather induced forces.
Cribbing	A box-like stack of inter-connected steel, timber, or concrete beams used to retain earth.
Crushed Rock	Gravel-size materials obtained by crushing larger rocks into smaller, angular

shape particles.

Cut
Removal of earth. Opposite of *fill*.

Deadman
Any kind of anchoring device set into the ground such as a block of concrete to anchor a lateral support.

Decomposed granite
A granular soil consisting of quartz, feldspar and mica. It results from the decomposition of granite.

Differential Settlement
In foundation design, the settlement of different parts of a building relative to each part. It is caused by varying compressibility of the underlying soil or by differing soil bearing pressures.

Direct Shear Test
In soil mechanics, a laboratory test to determine the strength parameters of soil.

Dowels:
Reinforcing steel placed in the footing and bent up into the stem a distance at least equal to the required development length.

Dry Density
The weight of the solid, dry, portion of soil or rock, expressed in pounds per cubic foot.

Earth Moving Equipment
Backhoe, bulldozer, clam shell, ditcher, drag line, grader, loader, shovel, scraper, skip loader and truck types.

Eccentricity
The distance from which a force acts away from the center of gravity of a member or connection, and which tends to cause such member to rotate or bend.

Elasticity
The degree to which a material can be stretched or shortened and returned to its original position when the applied force is removed.

Embankment
A term usually meaning a body of fill soil which has a length exceeding both width and height. Levy and highway fills are types of embankments.

Empirical
A term meaning to rely on observation or experimentation rather than upon theoretical analysis.

Erosion
The wearing away by water or wind.

Expansive Soil
Soil which changes in volume with changes in water content, swelling when wetted and shrinking when dried. Most clay soils are expansive. A soil is generally considered to be expansive if its volume change potential is enough to require a special foundation to support it such as piers or piles.

Expansive soils:
Consisting of clay, but some silt is also expansive. Some clays are highly expansive and change in volume with changes in water content.

Fault
In geology a fracture in the ground along which there is movement of one side relative to the other. The movement may be a few inches or several miles. Earthquakes are caused by stress relieving movements along faults.

Fill	Earth placed over natural ground. When fill material is brought to a site from another location it is called import fill. When material is removed from the site it is called export.
Flagstone	A type of stone which can be easily split into thin slabs in which is usually used for paving.
Floating Foundation	A raft-like concrete slab foundation which is used to distribute loads over a large area to reduce bearing pressure on the underlying soil.
Footings	Concrete placed on or into the soil to transfer loads from a structure to the soil. Footings are further classified as isolated footings such as a square foundation to support a column and continuous footings such as under the full length of a wall. Footings is a general term including both the continuous and spread footings which are bottomed near the ground surfaces as opposed to deep piles or caissons.
French Drain	A below-ground drain consisting with trench filled with gravel to permit movement of water through the gravel to a collection point.
Friction Piles	**See *Piles***
Frost line:	Term used in colder climates in the northern US, whereby upper portions of the ground may freeze seasonally or permanently, with depths ranging from a few inches to eight feet or more.
Geo Fabrics	Synthetic fabric used in or on the ground as a filtering material, or as a reinforcing membrane. It is available in rolls of either woven or non-woven sheets of synthetic polymers. Its use includes lining ditches and drains, protecting slopes from erosion and as an underlayment for pavements on poor soil.
Global Stability:	Similar to "slope stability", whereby an entire soil mass under one or two tiered retaining walls slips in a rotational pattern because of low shear strength of underlying soil. The walls remain intact with this type of failure, but the soil mass slips and rotate as a bowl-shaped mass.
Grade	The surface of the ground at a specific location.
Gradebeam	A continuous footing reinforced to act as a beam. It is used to bridge over localized weak soil, or to connect isolated spread footings.
Grading	In earth work, the leveling or otherwise shaping of the ground surface, usually by means of earth moving or excavation equipment. Rough grading is a processing of the ground to approximately the desired configuration usually within +/- 1/10 of a foot: Fine grading is the final contouring such as immediately before paving.
Grain Size Distribution	A plot (graph) of the percentage of each size of grain in the soil or aggregates

Curve	sample.
Granite	A type of igneous rock consisting of quartz, feldspar and mica. It is used in building construction because of its hardness and ability to take a polish.
Gravel	Rock particles which by convention range from ¼ to 3" in diameter. See also cobbles, sand, boulders.
Gravity retaining walls	This type of wall depends upon the dead load mass of the wall for stability rather than cantilevering from a foundation.
Grillage	A temporary means of support using tiers of short beams laid in alternating layers to form a box-like support which is usually temporary as under a house which has been jacked up ready to be moved.
Groundwater	Water standing or flowing between soil particles below the ground surface is caused primarily by a rainfall percolating into the soil and confined by dense soil a rock masses. The top surface is called a water table. If the ground water intersects the surface it forms a spring (if on a hill side) (or swamp if on level ground).
Gumbo	A highly plastic clay sticky when wet and which develops large shrinkage cracks upon drying.
Hard Pan	A layer of soil which has been naturally cemented by the leaching of minerals down from the soil strata above.
Hydrostatic Pressure	Pressure exerted by water.
Igneous Rock	One of the three classifications of rock. The other two being metamorphic rock and sedimentary rock. Formed within the earth by solidification of a molten mixture of minerals. Examples would include granite and basalt. Pumice is a volcanic ash.
Import Fill	In earth work, fill material which is brought to a site from another location, also called borrow.
In Situ Test	A test made within the soil at the site of origin: For example, testing in-place soil or rock.
Invert	The lowest point, usually referring to the flow line of a pipe or trench.
Limestone	A white to gray sedimentary rock which is chiefly composed of calcium carbonate. It is used as a source for lime.
Liquefaction	A temporary loss of strength condition in a saturated, loose, sandy soil caused by shock, such as an earthquake. It can cause serious settlement or failure of foundations.

Liquid Limit	In soil mechanics the water content at which soil pass from plastic to fluid state.
Loam	A soil classification meaning a mixture of equal parts of sand, clay and organic matter.
Loess	A homogeneous, non-layered loose soil deposited by wind action.
Mat Foundation	See *floating foundation*.
Maximum Density	In soil mechanics, the maximum amount of compaction that a soil can be given under optimum conditions of water content. For example: The term "90% compaction" means the soil is compacted to 90% of its maximum possible density determined by following a particular laboratory procedure.
Mechanical Analysis	A determination of soil or aggregate gradation by screening through a series of progressively smaller screens.
Metamorphic Rock	One of the three classifications of rock (the other two being an igneous and sedimentary rock) these are rocks which have changed from some previous structure due to pressure, heat, water or other agencies. For example, the change of limestone (a sedimentary rock) into marble.
Modulus of Elasticity	The ratio of stress to strain for an elastic material. By knowing the modulus of elasticity (symbol "E") of a material the amount of strain (deformation or movement) can be computed for a given of stress.
Modulus of Subgrade Reaction	A concept that states that the reactive pressure to a load on a soil, is proportional to the defection of the system that applies the load. A concept that is only marginally applicable to soil, but is otherwise widely used.
Overturning Moment	A structural design term meaning a horizontal force times a vertical distance, which tends to overturn a structural element such as a retaining wall, tank or comported of a building. The horizontal force can be wind, seismic or earth pressure. See also *resisting moment*.
Penetrometer	A device which measures the relative density and sometimes the bearing capacity of a soil by the resistance to penetration into soil of a cylindrical plunger or shaft. A Proctor needle is one type of a penetrometer.
Percolation Test	A test to determine the permeability of soil or rock by measuring the time and rate of change in the level of a column of water allowed to penetrate into the soil or rock.
Permafrost	A permanently frozen of soil.
Pile Cap	A concrete mass at the top of a group of piles to act as a means to transfer load from a structure into the piles.
Piles	Shaft-like structural members placed vertically into the soil to support a building or structure, as opposed to *Footings, Belled Piers,* or a *Floating*

Foundation. Piles support *Loads* by either the frictional resistance between the surface of the pile and the soil – called FRICTION PILES – or by *Column* action where the pile penetration is in poor soil and bears, as a column, upon a hard stratum at a deep level called BEARING PILES. Piles are inserted into the ground by several methods, including driving by repeated blows with a heavy weight, or *Pile Hammer*, vibrating with an oscillating *Pile Driver* attached to the upper end of the pile; *Jetting,* whereby high pressure water discharged ahead of the tip displaces the soil permitting easy penetration with only nominal driving; and placing *concrete* into predrilled bores – called CAST-IN-PLACE PILES. Types of piles include:

BATTER PILES – A pile placed into the ground at an angle, so as to provide increased resistance to *Lateral Loads* or *Uplift.*

CONCRETE PILES – These are either precast (made in a plant and shipped as a unit for driving at the job site) or cast-in-place as described above.

"H" PILES – Steel *'Beams,* which are available in a number of standard rolled shapes, which are driven into the ground by a pile driver.

STEP-TAPER FILE – This is a cast-in-place pile where successively reduced diameter *Steel* casings are drivien into the ground and subsequently filled with concrete in such a way that a "step-tapered" pile results.
WOOD PILES – Round wood piles driven into the ground. Those are usually tapered and are of *Pressure-Treated Wood* to impede deterioration.

Poisson's Ratio	A structure design, the ratio of sidewise elongation to lengthwise elongation for the material is subjected to stress.
Proctor Needle	In soil mechanics, a type of penetrometer having varying diameter tips which can be put on the plunger. It is used for measuring the relative density of soil by the resistance to penetration of the tip.
Quartz	The chief constituent of sand; a crystalline silicon dioxide material.
Quicksand	A phenomenon of sandy soil whereby the force of upward flowing water exceeds the downward weight to soil particles causing a suspension-like condition of the soil mass.
Rebar	Contraction of reinforcing bars.
Retaining Wall	Any wall intended to retain earth at a higher elevation on one side. For stability, it usually employs a wide footing. It should contain weep holes near the base to prevent hydrostatic pressure buildup. The corner of the footing under the retainer side is called the heel of the footing and the front corner, the toe.
Rip Rap	Rocks or boulders placed along the shoreline of a river or embankment to prevent erosion by water action.

Rock	A consolidated, hard, naturally forming mixture of materials, usually distinguished from soil by the difficulty of excavation. For example; igneous rock, metamorphic rock and sedimentary rock.
Sand	Mineral grains ranging in size from silty soil (.002 ") to gravel (0.2") in diameter. See also *Sandy Soil*.
Sandy Soil	A type of soil characterized by a predominance of sand-size particles. Sandy soil usually has low compressibility and high permeability.
Scouring	The lowering of the ground surface, such as around a foundation, by the erosive action of moving water or wind.
Sedimentary Rock	One of the three categories of rock (the others being igneous and metamorphic rock). It is formed from consolidation of fragments of rocks, minerals and organisms or deposits from sea water. Examples of sedimentary rock include sandstone and limestone.
Segmental retaining walls (SRWs)	Systems of stacked segmental concrete units, steel bins, or other devices that retain soil by tacking individual components. Most are patented systems that are typically battered (sloped backward), primarily to reduce lateral soil pressure, thus requiring a minimal foundation.
Settlement	The downward movement of a foundation due to consolidation of the underlying soil mass. Although some settlement always occurs, its magnitude is limited by design so undesirable distortion or cracking in the structure will not occur.
Sheeps Foot Roller	In earth work a heavy roller having many blunt ended protrusions which is pulled behind a tractor to compact filled ground.
Sheet Piling	A continuous, fence-like curtain of wood or interlocking steel sections which are driven into the ground to retain earth during excavation.
Shoring	A material used to provide temporary support of earth cuts during construction, or any temporary support.
Silica	The most common solid mineral known. In quartz or sand are largely silica. Chemically it is known as silica dioxide.
Silt	A fine grained unconsolidated soil with particles between sand and clay in size.
Skin Friction	In soil mechanics, the resistance to movement between the surface of a friction pile and the soil in contact with it.
Soil	An aggregation of loosely or firmly bound mineral grains, organic material, water and gas, resulting from the mechanical and chemical decomposition or dissolution of rock. Soil is differentiated from rock by its ability to be readily excavated by mechanical processes.

Soil Mechanics	The application of laws of mechanics and hydraulics to soils.
Soil Stabilization	Any artificial method used for improving soil bearing value, such as by compaction, the addition of add-mixtures, or preloading the soil by adding a temporary surcharge.
Soil Testing Laboratory	A laboratory especially equipped to test soil samples, usually shear strength, degree of compaction, expansiveness and other characteristics necessary for design of foundations.
Soldier Piles	Piles driven at regular intervals between which timbers are placed for temporary or permanent retainment of earth materials.
Specific Gravity	The ratio of the weight of a material to the weight of the same volume of water.
Statics	A fundamental concept of structural design having three basic laws:

1. The summation of all horizontal forces acting on a framework or structural member must equal zero.
2. The summation of all vertical forces acting on a structural member must equal zero and lastly . . .
3. The rotation causing moments about any point must equal zero. Ideally all of the above would include factor-of-safety (fos) for real-wall situations.

Subsidence	Downward movement of the ground surface due to consolidation of a subsurface zone. Subsidence may be caused by oil or water withdrawal, decay of organic matter in the subsoil, or similar phenomenon.
Subsoil	The soil zone below the topsoil.
Triaxial Compression Test	A test used primarily for determining the strength parameters of a soil under different conditions of consolidations and drainage. Carefully controlled loads are applied vertically and laterally to a soil sample until failure occurs.
Underpinning	Improving an existing foundation, such as to provide for an added load, or by constructing a new and larger replacement foundation.
Undisturbed Soil Sample	A sample of soil withdrawn from the ground as a core, usually by a sampling spoon, so that the soil can be examined or tested in its original state.
Water Table	The upper surface of a body of groundwater.
Weep holes:	Holes provided at the base of the stem for drainage. Weep holes usually have gravel or crushed rock behind the openings to act as a sieve and prevent clogging. Poor drainage of weep holes is the result of weep holes becoming clogged with weeds, thereby increasing the lateral pressure against the wall.

INDEX

A

AASHTO, 24
Active soil pressure, 18
Adhesion, piles and piers, 76
Adjacent footing surcharges, 30
Allowable Stress Design (ASD), 55
Anchored walls, 4
Angle of internal friction, 17
Angle of repose, 22
At rest soil pressure, 20
At-rest soil pressure, 117
Axial loads, 35

B

Backfill, 131
Bearing pressure, soil, 61
Bearing pressure, soil, 62
Boussinesq Method, 30
Bridge abutments, 4
Building Codes, 23
Bulkhead walls, 4
Buttress walls, 3

C

Caissons, 73
California Building Code, 23
Cascading Walls, 34
Clay soil, 14
Coefficient of friction, 52
Cohesion, 76
Cohesion resistance, 65
Cohesive soil, 14
Compaction, 132
Control joints, 131
Coulomb equation, 25
Counterfort walls, 2, 81
Crib walls, 3

D

d, 54
Definitions, 5
Deflection of Walls, 66
Design Checklist, 7
Detention walls, 34
Disclaimer, i
Dowels, 49
Drainage, 57, 131

E

Earthquake design, 37

Equivalent Fluid Pressure (EFP), 18
Expansive soil, 14, 15

F

Failures, 126
Flexural Strength, 71
Flood Walls, 34
Footing design, 69
Footing keys, 65
Footing Reinforcingt, 71
Footing shear, 71
Forensic Investigations, 132
Foundation investigations, 15
Freestanding walls, 115
Friction resistance, 65
Frost line, 14

G

Gabion walls, 3, 93
Geogrid Wall Design, 102
Geogrids, 103
Global stability, 67
Gravity walls, 91

H

Heel Reinforcing, 71
Highway surcharges, 30
Hooked bar embedments, 50
Hydrostatic Pressure, 34

I

Impact loading, 36
Inspection, 57
International Building Code, 23

K

K_{AE}, 43
Keys, 65
Keyway, 5, 52

L

Lagging design, 87
Lap splices, 55
Large block walls, 95
Lateral earth pressure, 28
Lateral earth pressures
    SRWs, 97
Lateral Earth Pressures, 25
Lateral soil pressure, 35, 105

Adjacent footing surcharges, 30
Boussinesq Method, 30
Coulomb equation, 25
Highway surcharges, 30
Impact Loading, 36
Mixed Soils, 29
Segmental walls, 111
Lateral wall pressure
Rankine equation, 25
Load Combinations, 11

M

Masonry pilaster walls, 115
Maximum Considered Earthquake (MCE), 41
Mercalli Scale, 38
Meyerhof Method, 64

N

NAVFAC, 24, 32
NEHRP, 24
NFPA, 23
Non-Yielding, 46
Non-Yielding Walls, 118

O

Overturning moments, 59

P

Passive resistance, 20, 65
Pickle Jar Test, 22
Pier design, 75
Piers, 73
Pilaster walls, 115
Pile Design, 74
Piles, 73
Precision Illusion, 22

R

Rankine equation, 25
Reinforced Earth, 102
Reinforcing, 52
Cover requirements, 55, 56
Footing, 69
Hooked embedments, 50
Horizontal, 51
Maximum, concrete, 56
Maximum, masonry, 54
Minimum, concrete, 56
Minimum, masonry, 54
Shrinkage, 72
Splices, 55, 57
Stem reinforcing, 69
Reinforcing:, 54
Resisting Moments, 60

Restrained walls, 46, 118
Richter Scale, 38
Rupture plane, 17

S

Safety Factors, 11
S_{DS}, 42
Seed and Whitman, 46
segmental retaining walls, 111
Segmental retaining walls, 98
Segmental walls, 3
Seismic, 44
Seismic, 46
Seismic design, 39, 110, 118
Restrained walls, 46, 118
segmental walls, 111
Segmental walls, 97
Stem self-weight, 46
Self-weight, 46
Shear strength, soil, 14
Sheet pile walls, 4, 120
Shrinkage reinforcing, 72
Skin friction, 76
Sliding, SRWs, 109
Sloping backfills, 19
Soil wedge, 16
Soil, mixed types, 29
Soils, 13
Soil , mixed types, 29
Soil Bearing Pressure, 62
Soil bearing values, 20
Soil bearing, SRW's, 109
Soil bearing, ultimate, 63
Soil density, 20
Soil designations, 13
Soil modulus, 21, 66
Soil pressure, 63
Soil, cohesive, 14
Soil, shear strength, 14
Soldier beam walls, 121
Special inspection, 57
SRWs, 97
Stability, 64
Stem Design, 49
Stem design, seismic, 41
Seismic, 44
Strength Design, 55
Surcharge Loads, 29
Surcharges, adjacent footings
NAVPAC, 32
Surcharges, highway, 30
Swimming pool walls, 113

T

Temperature reinforcing, 72
Terminology, 1, 6, 239
Terzaghi and Peck Method, 32
Terzaghi, Karl, 13
The Failure plane, 26

The Mononobe-Okabe equation, 41
Tie-back design, 117, 119
Tilt, 21, 66
Tilt of Walls, 66
Tilt-Up retaining walls, 85
Tilt-up walls, 3
Toe Reinforcing, 70
Torsion, 79
Types of Retaining Structures, 1

U

Uniform Building Code, 23
Uniform System of Soil for Classification, 13

V

Vertical Component, 61
Vertical loads, 35
Vertical loads, axial, 35

W

Wall friction angle, 25, 26
Water Table Conditions, 33
Weep holes, 5
Weight, soil, 16
Whitney Approximation Method, 78
Wind forces, 33
Wood post design, 88
Wood retaining walls, 3, 87
Wythe definition, 93

NOTES